The Wonders of Physics

3rd Edition

T0350054

The Wonders of Physics

3rd Edition

Andrey Varlamov
Italian National Research Council, Italy

Lev Aslamazov
Moscow Technological University, Russia

Scientific Editor
A.A. Abrikosov, jr.

Translators
A.A. Abrikosov, jr., J. Vydryg, & D. Znamenski

World Scientific

NEW JERSEY · LONDON · SINGAPORE · BEIJING · SHANGHAI · HONG KONG · TAIPEI · CHENNAI

Published by

World Scientific Publishing Co. Pte. Ltd.
5 Toh Tuck Link, Singapore 596224
USA office: 27 Warren Street, Suite 401-402, Hackensack, NJ 07601
UK office: 57 Shelton Street, Covent Garden, London WC2H 9HE

British Library Cataloguing-in-Publication Data
A catalogue record for this book is available from the British Library.

THE WONDERS OF PHYSICS
(3rd Edition)

ISBN-13 978-981-4374-15-6
ISBN-10 981-4374-15-6

Printed in Singapore by Mainland Press Pte Ltd.

To our teachers and friends. . .

Preface

It is my pleasure to present the third English edition of *The Wonders of Physics*. A quarter of a century has passed since its first publication in Russia. The book had a long life; it was well accepted by readers. After the first release the second, third and fourth Russian editions followed, then two versions in English, one in Italian, and one in Spanish.

The world has undergone considerable changes during the last quarter of a century. Our vital priorities have changed. A new generation of readers grew up; new achievements in science and technology the authors could not even dream about while working on the first edition in the mid-1980s, became a part of our life. Side by side with the progress occurring in physics this book was undergoing constant development. It became twice as long, and new chapters appeared in it, telling readers about such fundamental discoveries as high-temperature superconductivity, nuclear magnetic resonance, a new vast area of modern research called nanophysics, and, naturally, about applications of new discoveries in everyday life.

For the last ten years the author (AV) has been living in Italy, where the culture of food is one of the priorities among the public interests. My first publications in the Italian press about the physics of coffee and physical aspects of enology came as a result of my efforts to understand how the laws of physics work in my newly discovered field of everyday human activity. These works, unexpectedly, found widespread interest and positive response; they were translated into several languages and discussed on TV. This is the reason why a new part about "gastronomic physics" appeared in the Russian edition of the book. I hope our readers will enjoy it.

The reasons Lev Aslamasov (1944 - 1986) and I wrote this book were to satisfy both our curiosity and our wish to share our admiration for the beauty of physics in all its natural manifestations. We devoted a lot of time to teaching physics at various levels, from gifted freshmen to mature Ph.D. students. All our experience has convinced us that, in addition to obvious necessity of regular and rigorous study of this discipline, an "artistic" approach, in which the teacher (or the author) proves the importance of physics in everyday phenomena, is vital. I hope we have managed to communicate our perception of physics not only in the title but also in the body of the book. The book is based on articles published in the *Kvant* Magazine and other journals during the last 40 years.

I would like to express my profound gratitude to many friends and colleagues, without whom this edition would not be possible. First, a science editor and my old and dear friend Dr. Alex Abrikosov (Jr.), whose enthusiasm, thorough scientific expertise, corrections, comments and translation gave birth to the English version. His contribution to the project was considerably enforced by the collaboration with my friends Dr. Dmitriy Znamensky, and Janine Vydrug. Several chapters of this book were written based on joint publications with my friends: Professor G. Balestrino, Professor A. Buzdin, Professor Yu. Galperin, Professor A. Rigamonti, and my high school teacher Dr. A. Shapiro. I would like to acknowledge their contribution.

Finally I would like to thank warmly my editors: Dr. D. De Bona, Dr. L. Panyushkina, Dr. T. Petrova, Dr. V. Tikhomirova, and Dr. A. Ovchinnikov without whose professionalism and collaboration in the preparation of previous editions, the present one would not appear.

In conclusion, I would like to cordially mention on behalf of myself and Alex Abrikosov (Jr.) three more people. Two of them are Alex's parents and our teachers of physics and life, Alexei and Tatyana Abrikosov. The third is our mutual friend from our student days, Serguei Pokrovski. These people played a vital part in our formation.

A. A. Varlamov,
(Rome, 2011).

From the foreword to the Russian edition

The science of physics was at the head of scientific and technical revolutions in the twentieth century. Nowadays physics continues to determine the direction of human progress. The biggest example is that of the recent discovery of high-temperature superconductivity, which may radically change the entire edifice of modern technology.

However, delving deeper into the mysteries of the macrocosm and microparticles, scientists move further away from traditional school physics, that treats transformers and projectiles, launched at an angle to the horizontal, namely, from what most people believe to be physics. The goal of popular literature is to bridge the gap, and to bring the excellence of modern physics to curious readers while demonstrating its major achievements. A difficult task that won't tolerate dabbling.

The book in your hands develops the best traditions of this kind of literature. Written by working theoretical physicists and, at the same time, the dedicated popularizers of scientific knowledge, clear and captivating in manner, it brings to the reader the latest achievements in quantum solid-state physics; but on the way it shows how the laws of physics reveal themselves even in seemingly trivial episodes and natural phenomena around us. And most importantly, it shows the world through the eyes of scientists, capable of "proving harmony with algebra".

It was a great loss that one of the authors of the book, the well-known specialist in the theory of superconductivity, professor L. G. Aslamazov, who for a long time was the vice-editor of the popular magazine *Kvant*, did not live to see the book come out.

I hope that readers, ranging from high-school students to professional physicists, will find this book, marked by its extremely vast scope of encompassed questions, a really interesting, enjoyable and rewarding read.

A. A. Abrikosov, 2003 Nobel Prize winner,
(Moscow, 1987).

Translator's note

The offer to translate this book into English was a great honor for me. Now I'm your interpreter in the marvelous land of physics. But this is not a simple coincidence.

First of all, for me physics is a sort of "family business"; that you no doubt might have guessed. Many of the people I remember warmly from my first days, turned out to be physicists. As a ten-year-old schoolboy I remember (then postgraduate, later professor) Lev Aslamazov sunbathing on the Odessa beach.[a] Then, in high-school, I met my best friend Andrei Varlamov. We made our decision and both entered the Moscow Institute of Physics and Technology. For long hours we discussed and argued about many things related and not related to physics. Some of the topics in this book awaken memories of those days.

My "physical orientation" would not be complete without the newly established Moscow magazine *Kvant,* with its enthusiastic young team of editors. (Its English translation is *Quantum.*) L. Aslamazov was at the very origin of it and his article "Meandering down to the sea" appeared in the first issue. As we grew older we started writing ourselves. And almost every chapter of this book once has appeared under *Kvant* cover.

Somewhere among my papers is the draft of my first popular paper. Leva (as everybody called him) rejected it, explaining that we must not simply write about what we knew from textbooks, but find new bright and clear illustrations of our knowledge. According to him, this was the main and most difficult task in popularization of science. And, as you shall see, this is the spirit of the present book.

My love of physics swung the balance in favor of translating the book although I'm not a native English speaker. I hope that a share of nonlinguistic knowledge that I have tried to invest in the text will, at least partially, compensate for its "Russian flavor", and you will be amused rather than annoyed by some inevitable slips.

Sure enough I would not take the risk alone and what you read is a result of real collaboration with my fellow translator Dmitriy Znamensky from whom I learned so much. Writing this note myself is only the privilege of an old acquaintance and this must by no means belittle his contribution.

---

[a]Odessa is the city on the Black Sea coast where traditional spring symposia on theoretical physics were held.

In completing the translation, we have tried to commemorate the great scientists of the past and have supplied the text with short biographical footnotes to this effect.

A. A. Abrikosov, Jr., (—A. A.).

(Moscow, 2000).

Contents

Fig. 1. Schematic image of wave.

Fig. 2. Dependence of human eye perception on wavelength of light.

Fig. 3. Arkadii Rylov "In the blue".

PART 1

Outdoor physics

From the first part of the book the reader will learn why rivers wind and how they wash their banks out, why the sky is blue and white horses are white. We are going to tell you about the properties of the ocean, about winds and about the role of the Earth's rotation.

In short we shall present examples of how the laws of physics work on a global scale.

Chapter 1

Meandering down to the sea

Have you ever seen a straight river without bends? Of course a short section
of a river may cut straight but no rivers are entirely without bends. Even
if the river flows through a plain, it usually loops around and the bends
repeat periodically. Moreover, as a rule, one bank at the bend is steep while
the other slopes gently. How could one explain these peculiarities of river
behavior?

Hydrodynamics is the branch of physics that deals with the motion
of liquids. Although it's a well-developed science, rivers are complicated
natural objects and even hydrodynamics can't explain every feature of their
behavior. Nevertheless, it can answer many questions.

You may be surprised to learn that even the great Albert Einstein[a]
gave time to the problem of meanders. In the report delivered in 1926 at a
meeting of the Prussian Academy of Sciences, he compared the motion of
river water to swirling water in a glass. The analogy allowed him to explain
why rivers choose twisted paths.

Let's try to understand this too, at least qualitatively. And let's start
with a glass of tea.

1.1 Tea-leaves in a glass

Make a glass of tea with loose tea-leaves (no tea-bags!), stir it well, and take
the spoon out. The brew will gradually stop and the tea-leaves will gather

---

[a] A. Einstein, (1879–1955), German-born physicist, lived in Switzerland, US citizen from
1940; creator of the theory of relativity; Nobel Prize 1921 in physics.

in the center of the bottom. What made them settle there? To answer this question let us first determine what shape the surface takes as the liquid swirls in the glass.

The tea-cup experiment shows that the surface of the tea becomes curved. The reason is clear. In order to make particles of water move circularly, the net force acting on each of them must provide a centripetal acceleration. Consider a cube situated in the liquid at a distance r from the axis of rotation, Figure 1.1, a. Let the mass of tea in it be Δm. If the angular speed of rotation is ω then the centripetal acceleration of the cube is $\omega^2 r$. It comes as the result of the difference of the pressures acting onto the faces of the cube (the left and right faces in Figure 1.1 a). So,

$$m \, \omega^2 \, r = F_1 - F_2 = (P_1 - P_2) \, \Delta S, \qquad (1.1)$$

where ΔS is the area of the face. The pressures P_1 and P_2 are determined by the distances h_1 and h_2 from the surface of the liquid:

$$P_1 = \rho \, g \, h_1 \qquad \text{and} \qquad P_2 = \rho \, g \, h_2, \qquad (1.2)$$

where ρ is the density of the liquid and g is the free fall acceleration. As soon as the force F_1 is greater than F_2, h_1 must exceed h_2 and the surface of the rotating liquid must be curved, as shown in Figure 1.1. The faster the rotation is, the greater is the curvature of the surface.

One can find the shape of the curved surface of the revolving liquid. It turns out to be a paraboloid — that is, a surface with a parabolic cross section[b].

As long as we continue stirring the tea with the spoon, we keep it swirling. But when we remove the spoon the viscous friction between layers of the liquid and the friction against the walls and bottom of the glass will convert the kinetic energy of liquid into heat, and the motion will gradually come to rest.

As the rotation slows down, the surface of the liquid flattens. In the mean time vortex currents directed as shown in Figure 1.1, b appear in the liquid. The vortex currents are formed because of the nonuniform deceleration of the liquid at the bottom of the glass and at the surface. Near the bottom, where the friction is stronger, the liquid slows down more effectively than at the surface. So, despite being at equal distances from

---

[b]The form of the surface is parabolic only if the liquid is rotated together with the glass as a whole. This is called *rigid rotation*. —A. A.

Fig. 1.1: *a* --- Hydrostatic forces acting on a particle in rotating liquid. *b* --- Vortex currents arising as rotation slows down.

the axis of rotation particles of liquid acquire different speeds: the ones that are closer to the bottom become slower than those near the surface. However the net force due to the pressure differences is the same for all these particles. This force can't cause the required centripetal acceleration of all the particles at once (as was the case with uniform rotation with the same angular speed). Near the surface, the angular speed is too large, and particles of water are thrown to the sides of the glass; near the bottom the angular speed is too low, and the resultant force makes water move to the center.

Now it is clear why tea-leaves gather in the middle of the bottom, Figure 1.2. They are drawn there by vortex currents that arise due to the nonuniform deceleration. Of course, our analysis is simplified but it accurately grasps the main points.

1.2 How river-beds change

Let's consider the motion of water in a river bend. The picture resembles what we have observed in our glass of tea. The surface of water inclines inside the bend in order that pressure differences produce the necessary

Fig. 1.2: The tea-cup experiment. Vortex currents
drive tea-leaves to the center of the bottom.

centripetal accelerations. (Figure 1.3 shows schematically a cross section
of bending river.) Quite similarly to the tea glass, the velocity of water
near the bottom is less than that near the surface of the river (distribution
of velocities with depth is shown by vectors in Figure 1.3). Near the
surface the net difference in hydrostatic pressures can't make the faster
water particles follow the curve and the water is "thrown" to the outer shore
(away from the center of the bend). Near the bottom, on the other hand, the
velocity is small, so the water moves toward the inner shore (to the center
of the bend). Hence additional circulation of water appears in addition to
the main flow. Figure 1.3 shows the direction of water circulation in the
transverse plane.

Fig. 1.3: Cross
section of a turning
river-bed.
Hydrostatic forces,
vortex currents and
velocity distribution.

The circulation of water causes soil erosion. As a result, the outer bank
is undermined and washed out while the soil gradually settles along the

inner shore, forming an ever thickening layer (remember the tea-leaves in the glass!). The shape of the river-bed changes so that the cross section resembles the one shown in Figure 1.4. It's also interesting to observe how the velocity of the stream varies across the river (from bank to bank). In straight stretches water runs most quickly in the middle of the river. At bends, the line of fastest flow shifts outwards. This happens because it's more difficult to turn fast-moving water particles than slow-moving ones. A larger centripetal acceleration is necessary. But the greater the velocity of the flow the circulation of the water, and consequently, the greater the soil erosion. That's why the fastest place in a river-bed is usually the deepest one — a fact well known by river pilots.

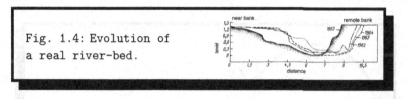

Fig. 1.4: Evolution of a real river-bed.

Soil erosion along the outer bank and sedimentation along the inner one result in a gradual shift of the entire river-bed away from the center of the bend thereby increasing the river meander. Figure 1.4 shows the very same cross section of a real river-bed at several year intervals. You can clearly notice the shift of the river-bed and the increase in its meander.

So even an occasional slight river bend — created, for example, by a landslide or by a fallen tree — will grow. Straight flow of a river across a plain is unstable.

1.3 How meanders are formed

The shape of a river-bed is largely determined by the relief of the terrain it crosses. A river passing a hilly landscape winds in order to avoid heights and follows valleys. It "looks for" a path with the maximum slope.

But how do rivers flow in open country? How does the instability of a straight river-bed with respect to bending described above influence the course of a stream? The instability must increase the length of the path and make the river wind. It's natural to think that in the ideal case (an absolutely flat, homogeneous terrain), a periodic curve must appear. What will it look like?

Geologists have put forth the idea that at their turns, paths of rivers flowing through plains should take the form of a bent ruler.

Take a steel ruler and bring its ends together. The ruler will bend like it is shown in Figure 1.5. This special form of elastic curve is called the *Euler curve* after the great mathematician Leonard Euler[c] who analyzed it theoretically. The shape of the bent ruler has a wonderful property: of all possible curves of a fixed length connecting two given points, it has the minimum average curvature. If we measure the angular deflection θ_k (Figure 1.5) at equal intervals along the curve and add up their squares then the sum $\theta_1^2 + \theta_2^2 + \ldots$ will be minimal for the Euler curve. This "economic" feature of the Euler curve was basic for the river-bed shape hypothesis.

Fig. 1.5: The form of a bent steel ruler is called the Euler curve.

To test this hypothesis geologists modeled a changing river-bed. They passed water through an artificial channel in a lightly erodible homogeneous medium composed of small, weakly-held-together particles. Soon the straight channel began to wander, and the shape of the bend was described by the Euler curve (Figure 1.6). Of course, nobody has ever seen such a perfect river-bed in nature (because of the heterogeneity of the soil, for instance). But rivers flowing through plains usually do meander and form

[c]L. Euler, (1707–1783), Swiss-born mathematician and physicist; member of Berlin, Paris, St. Petersburg academies and of the London Royal Society; worked a long time in Russia. Buried in St. Petersburg.

periodic structures. In Figure 1.6 you can see a real river-bed and the Euler curve (the dashed line) that approximates its shape best of all.

Fig. 1.6: *a* --- Modeling meanders in laboratory. Channel-bed develops periodic Euler bends (the dashed line). *b* --- A real river and the nearest Euler curve (the dashed line).

By the way, the word "meander" itself is of ancient origin. It comes from the *Meander*, a river in Turkey famous for its twists and turns (now called the *Menderes*). Periodic deflections of ocean currents and of brooks that form on surfaces of glaciers are also called meanders. In each of these cases, random processes in a homogeneous medium give rise to periodic structures; and though the reasons that bring meanders about may differ, the shape of the resulting periodic curves is always the same.

Show that the surface of a uniformly (rigidly) revolving liquid takes a parabolic form.

Chapter 2

Rivers from lakes

...Old Baikal had more than three hundred sons and
the sole daughter, beautiful Angara.

Ancient legend.

Let skeptical readers disbelieve the epigraph and search geographical at-
lases. Jokes aside, he will discover that 336 rivers fall into the Lake Baikal
and only one, the mighty Angara, has it for a source. It turns out that
Baikal is not the exception. As a rule, no matter how many rivers fall into
a lake, only one comes out of it.

For example many rivers run to Ladoga lake but only the Neva escapes
it; Svir is the only outflow of the Onega lake, etc. This may be explained by
the fact that the outflowing water prefers the deepest river-bed and other
possible exits are left above the level of the lake. It is hardly probable that
openings of several would-be river-beds have the same elevation. A copious
water supply in a brimming lake can send forth two streams. However
such a situation is not steady and may take place only in relatively young
(recently formed) lakes. Little-by-little, the deeper and faster stream will
wash away the bed increasing the outflow. As a result the level of the lake
will decline and the weaker flow will be gradually silted. Thus only the
deepest of the outflowing rivers will survive.

In order for a lake to be the source of two rivers it is necessary that
their origins lay exactly at the same level. This case is called a *bifurcation*
(the term is now widely used by mathematicians to indicate the increase of
the number of solutions to an equation). Bifurcations are uncommon and
usually only a single river comes out of a lake.

The same laws may be applied to rivers. It is well known that rivers readily flow together, whereas forks are relatively infrequent. Streams always prefer the steepest descending curve. The probability of splitting of this curve is small. Nevertheless the situation changes at the river delta where the main stream divides into many smaller channels.

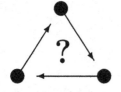

Try to figure out why rivers behave so strangely when approaching a big lake or the sea.

Chapter 3

The oceanic phone booth

The walls, indeed, had ears. Or, rather, an ear. A round hole with a tube – a sort of secret telephone – conveyed every word said in the dungeon right to the chamber of Signore Tomato.

Gianni Rodari, *"The Adventures of Cipollino."*

Not so long ago – in the mid-1940s, to put a number on it – scientists from the USSR and US discovered an amazing phenomenon. Sound waves propagating in the ocean could sometimes be detected thousands of kilometers away from the source. In one of the most successful experiments, the sound from an underground explosion set off by scientists on the coast of Australia traveled halfway around the globe and was registered by another group of researchers in Bermuda, some 19,600 km away (a record distance for the propagation of pulse sound signals). This means that the intensity of the sound didn't change greatly as it ran away from the source. What is the mechanism for such long-distance propagation of sound?

It looked as if the ocean contained an acoustic waveguide — that is, a channel along which sound waves traveled practically without attenuation (loss of strength). You have read about such a means in the epigraph.

Another example of an acoustic waveguide is the tube used on ships from time immemorial. The ship's captain uses a tube to give orders to the engine room from the bridge. It's interesting that the attenuation of sound traveling in air along a waveguide is so small that if we constructed a tube 750 km long, it could serve as a "telephone" for calls between Pittsburgh and Detroit. But it would be inconvenient to chat over such a line, because your friend at the other end would have to wait half an hour to hear your words.

We should emphasize that the reflection of a wave from a waveguide's boundaries is a crucial feature of the waveguide: it's because of this very property that the wave energy doesn't scatter in all directions but follows the given one.

These examples would lead us to suppose that the propagation of sound over extremely large distances in the ocean is due to some sort of waveguide mechanism. But how is such a gigantic waveguide formed? Under what conditions does it appear, and what are the reflective boundaries that make sound waves travel so far?

Since the ocean surface can reflect sound fairly well, it might serve as the upper boundary of the waveguide. The ratio of the intensity of the reflected wave to that of a wave penetrating the interface between two media strongly depends on the densities and the speeds of sound in them. If the media differ substantially then even the sound falling normally onto a flat interface will be practically completely reflected. The densities of air and water differ a thousand times, the ratio of sound velocities in them is 4.5. Therefore the intensity of the normal wave passing to air from water is only 0.01% of the intensity of the incident sound. The reflection is even stronger when the wave falls onto the interface obliquely. But, of course, the ocean surface can't be perfectly flat because of the ever-present waves. This causes chaotic reflections of sound waves and disturbs the waveguide character of their propagation.

The results aren't any better when the sound waves reflect off the ocean floor. The density of sediments at the bottom of the sea is usually within the range 1.24–2.0 g/cm^3, and the velocity of sound propagation in these sediments is only 2–3% less than that in water. So when a sound wave hits the bottom a significant amount of its energy is absorbed.

Since the ocean floor poorly reflects sound, it can't serve as the lower boundary of the waveguide. The boundaries of the oceanic waveguide must be sought somewhere between the floor and the surface. And that's where they were found. The boundaries turned out to be the water layers at certain depths in the ocean.

How do sound waves reflect from the "walls" of the oceanic acoustic waveguide? To answer this question we'll have to examine the mechanism of sound propagation in the ocean.

3.1   Sound in water

Up till now, as we've talked about waveguides, our unspoken assumption was that the speed of sound in them is constant. But the speed of sound in the ocean varies from $1.450\,m/s$ to $1.540\,m/s$ depending on temperature, salinity, hydrostatic pressure, and other factors. The increase in hydrostatic pressure $P(z)$ with depth z, for instance, adds to the speed of sound $1.6\,m/s$ per 100 meters down. An increase in temperature also adds to the speed of sound. However, the water temperature, as a rule, falls rapidly as one descends from the well-warmed upper layers to the ocean depths, where the temperature is practically constant. Due to the interplay of these two mechanisms — the hydrostatic pressure and the temperature — the dependence of the sound velocity $c(z)$ on depth in the ocean looks like that shown in Figure 3.1. Near the surface the rapidly dropping temperature takes the upper hand. Here the speed of sound decreases with depth. As we plunge deeper, the rate of decrease in temperature slows, but the hydrostatic pressure continues to grow. At some depth the two factors balance: the speed of sound reaches its minimum. Deeper down the sound velocity starts to grow due to the rise in hydrostatic pressure.

Fig. 3.1: Velocity of sound c in ocean water depends on depth z and has a minimum at z_m.

We see that the speed of sound in the ocean depends on depth and this influences the nature of sound propagation. To understand how "sound beams" travel in the ocean, we'll turn to an optical analogy. We'll examine how a light beam propagates in a stack of flat parallel plates with varying indices of refraction. Then we'll generalize our findings to a medium with a smoothly varying refraction index.

3.2 Light in water

Let's consider a pile of flat parallel plates with varying indices of refraction $n_0, n_1, \ldots n_k$, where $n_0 < n_1 < \ldots < n_k$ (see Figure 3.2). Assume that the light beam falls onto the upper plate at an angle α_0 relative to the normal. After refraction it leaves the 0–1 boundary at an angle α_1 and that is the incidence angle for the 1–2 boundary. Upon refracting at the next interface the beam hits the 2–3 boundary at an angle α_2 and so on. According to Snell's[a] law, we have:

$$\frac{\sin \alpha_0}{\sin \alpha_1} = \frac{n_1}{n_0}, \quad \frac{\sin \alpha_1}{\sin \alpha_2} = \frac{n_2}{n_1}, \ldots \quad \frac{\sin \alpha_{k-1}}{\sin \alpha_k} = \frac{n_k}{n_{k-1}}.$$

Remembering that the ratio of the refraction indices of two media is the inverse of the ratio of the speeds of light in these media, we may rewrite these equations in the following form:

$$\frac{\sin \alpha_0}{\sin \alpha_1} = \frac{c_0}{c_1}, \quad \frac{\sin \alpha_1}{\sin \alpha_2} = \frac{c_1}{c_2}, \ldots \quad \frac{\sin \alpha_{k-1}}{\sin \alpha_k} = \frac{c_{k-1}}{c_k}.$$

Multiplying the series of equations by one another, we get

$$\frac{\sin \alpha_0}{\sin \alpha_k} = \frac{c_0}{c_k}.$$

Reducing thickness of plates to zero while increasing their number to infinity, we approach the generalized law of refraction (Snell's law):

$$c(z) \sin \alpha(0) = c(0) \sin \alpha(z),$$

where $c(0)$ is the speed of light at the point where the beam enters the medium and $c(z)$ is the speed of light at a distance z from the boundary. Thus, as a light beam propagates through an optically nonuniform medium with decreasing index of refraction, it increasingly deflects from the normal line. As the speed of light in the medium decreases (and the index of refraction increases) the beam gradually turns parallel to the interface.

If we know how the speed of light varies in an optically nonuniform medium, we can use the Snell's law to find the trajectory of any beam in it. Sound beams propagating in an acoustically nonuniform medium, where the speed of sound varies, get deflected exactly in the same way. An example of such a medium is the ocean.

[a] W. Snell van Royen, (died 1626), Dutch mathematician.

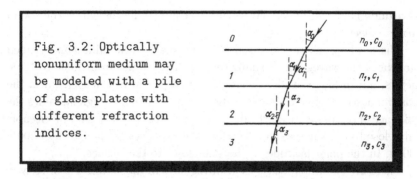

Fig. 3.2: Optically nonuniform medium may be modeled with a pile of glass plates with different refraction indices.

3.3 Water waveguides

Now let's get back to the question of sound propagation in the oceanic acoustic waveguide. Imagine that the sound source is located at the depth z_m corresponding to the minimum sound velocity (Figure 3.3). How do the sound beams travel as they leave the source? The beam propagating along a horizontal line is straight. But the beams leaving the source at an angle to the horizontal will bend because of sonic refraction. Since the speed of sound increases both up and down from the level z_m, sound beams will bend towards the horizontal. At a certain point the beam will get parallel to the horizontal and after being reflected it will turn back toward the line $z = z_m$, Figure 3.3.

Fig. 3.3: Refraction of sound in acoustically nonuniform medium (the sound velocity $c(z)$ is minimal at the plane $z = z_m$).

Thus, the refraction of sound in the ocean allows a portion of the sonic energy emitted by the source to propagate through the water without rising up to the surface or dropping down to the ocean floor. This means that we have an oceanic acoustic waveguide. The role of the "walls" of the waveguide is played by the layers of water at the depths where the sound beams reflect.

The level z_m where the speed of sound reaches the minimum is called the axis of the waveguide. Usually the depths z_m range from 1,000 to 1,200 meters, but in low latitudes, where the water is warmed deeper down, the axis can descend down to $2,000\ m$. On the other hand, in high latitudes the influence of temperature on the distribution of the sound speed is noticeable only close to the surface, and therefore the axis rises to the depth of $200 - 500\ m$. In the polar latitudes it lies still closer to the surface.

There are two different types of waveguides in the ocean. The first type occurs when the speed of sound near the surface (c_0) is less than that at the ocean floor (c_f). This usually occurs in deep waters, where the pressure at the bottom reaches hundreds of atmospheres. As we mentioned above, sound reflects well from the water-air interface. So if the ocean surface is smooth (dead calm), it can serve as the upper boundary of a waveguide. The channel then spreads through the entire layer of water, from the surface to the floor (see Figure 3.4).

Fig. 3.4: Acoustic waveguide of the first type: dead calm, $c_f > c_0$. Sound is reflected from the surface and refracted at the bottom.

Let's see what portion of the sound beam is "captured" by the channel. We start by rewriting Snell's law as

$$c(z)\cos\varphi_1 = c_1 \cos\varphi(z).$$

where φ_1 and $\varphi(z)$ are the angles formed by sound rays with the horizontal at depths z_1 and z, respectively. It's clear that $\varphi_1 = \frac{\pi}{2} - \alpha_1$, $\varphi(z) = \frac{\pi}{2} - \alpha(z)$. If the source of sound is located on the axis of the channel ($c_1 = c_m$) then the last sound ray captured by the channel is the one tangent to the ocean floor, $\varphi(z) = 0$, as shown in Figure 3.4. Therefore all rays that leave the source at angles satisfying the condition

$$\cos \varphi_1 \geq \frac{c_m}{c_f},$$

enter the channel.

When the water surface is rough, all sound beams will scatter from it. The rays leaving the surface at angles larger than φ_1 will reach the floor and be absorbed there. Yet even in this case, thanks to refraction, the channel can capture those rays that do not quite reach the rough surface (Figure 3.5). Then the channel spreads from the surface to a depth z_k which can be determined from the condition $c(z_k) = c_0$. It's clear that such a channel captures all sound rays with angles

$$\varphi_1 \leq \arccos \frac{c_m}{c_0}.$$

Fig. 3.5: Another acoustic waveguide of the first type: rough sea, $c_f > c_0$. Sound is refracted under the surface but does not reach the bottom.

The second type of waveguide is characteristic of shallow water. It occurs only when the speed of sound near the surface is greater than that at the floor, as in Figure 3.6. The channel occupies the water layer from the ocean floor up to the depth z_k where $c(z_k) = c_f$. It looks like a waveguide of the first type flipped upside down.

For certain types of dependencies of sound speed on depth, the waveguide foci sound beams like a lens. If a sound source is located on the axis, the rays that leave it at different angles will periodically converge at some points on the axis. These points are called foci of the channel. So if the dependence of the speed of sound in the channel on depth is close to parabolic: $c(z) = c_m \left(1 + \frac{1}{2} b^2 \Delta z^2\right)$, where $\Delta z = z - z_m$. Then for rays leaving the

Fig. 3.6: Acoustic waveguide of the second type. When $c_f < c_0$ the sound refracted at the bottom does not reach the surface.

source at small angles with the horizontal, the foci will lie at the points $x_n = x_0 + \pi n / b$, where $n = 1, 2, \ldots$ and b is a coefficient whose dimension is inverse to depth (m^{-1}), Figure ??. The parabolic function $c(z)$ is close to the actual dependence of the speed of sound on depth in deep oceanic acoustic waveguides. Deviations of $c(z)$ from parabolic law blurs the foci on the axis of the waveguide[b].

Fig. 3.7: Waveguide foci sound beams from the source (x_0) like an acoustic lens. Beams cross at points x_n called foci.

[b]Like many periodic processes in nature, the propagation of beams along a parabolic waveguide obeys the harmonic law. Near the axis trajectories follow the equation:

$$\frac{d^2 \, \Delta z}{dx^2} = -b^2 \, \Delta z,$$

where x is not the time but the horizontal coordinate. Obviously the trajectories are sinusoidal, $\Delta z = A \sin b \, (x - x_0)$, cross at the zeros of the sine, $x_n - x_0 = \pi n / b$.
—A. A.

3.4 Applications?

Is it possible to send a sound signal along an oceanic acoustic waveguide and receive it at the point of origin, after it has completely circled the globe? The answer is a flat no. First and foremost, the continents present insurmountable obstacles, as well as great contrasts in depths of the world's oceans. So it's impossible to choose a direction along which there would be a continuous round-the-world waveguide. But that isn't the only reason. A sound wave propagating along an oceanic acoustic waveguide differs from sound waves in the "telephone" tubes on ships that we mentioned at the outset. The sound wave traveling from the bridge to the engine room is one-dimensional, and the area of its wave front is constant at any distance from the source. Therefore, the strength of the sound will also be constant everywhere along the tube (heat losses aren't taken into account). As for the oceanic acoustic waveguide, the sound wave doesn't propagate along a straight line but in all directions in the plane $z = z_m$. So the wave here is a cylindrical surface. Because of this, the strength of the sound decreases with distance – that is, the sound intensity is proportional to $1/R$, where R is the distance from the source of the sound to the detector. (Try to obtain this dependence and compare it with the law of attenuation of a spherical sound wave in three-dimensional space.)

Another reason of sound attenuation is the damping of the sound wave as it travels through the waters of the ocean. Energy from the wave is transformed into heat due to viscosity of water and other irreversible processes. Moreover, sound waves dissipate in the ocean because of various heterogeneities, such as suspended particles, air bubbles, plankton, and even the swim-bladders of fish.

Finally we should point out that the underwater sound channel isn't the only example of waveguides in nature. Long-distance broadcasting from radio stations is possible only because of the propagation of radio waves through the atmosphere along giant waveguides. And we're sure you've heard of mirages, even if you've never seen one. Under certain atmospheric conditions, waveguide channels for electromagnetic waves in the visible range can form. This explains the appearance of a ship in the middle of desert, or a city that springs to life in the middle of the ocean.

Prove that the trajectories of sound beams in a parabolic waveguide obey the harmonic equation.

Chapter 4

In the blue

Now the wild white horses play,
Champ and chafe and toss in the spray.

Matthew Arnold, *"The Forsaken Merman."*

Artists are endowed with sharp professional vision. This makes the world depicted in realistic landscapes outstandingly bright and colorful and features some natural events. Even the best painters do not need to understand the reasons of the portrayed processes which may often be hidden deeply. However with the help of the landscapes of a good painter one can study ambient nature even better than in the physical world. Suspending the moment in the picture the author intuitively focuses on the principals and omits inessential and passing details.

Look at the picture "In the Blue" by the Russian painter Arkadii Rylov[a] that is reproduced on the color plate at the front of the book. "White birds soar among white clouds bathing in the blue sky. And the ship under the sails drifts peacefully over the gently rolling blue waves like a white bird in the ocean." This description belongs to the notable Russian art critic A. A. Fedorov-Davydov. Watching the wonderful canvas in the Moscow Tretyakov Gallery one forgets being in a museum and feels like a guest at this feast of nature.

But let us leave the aesthetic side and examine the picture with the eyes of researchers. First of all, from where did the author paint the landscape? Was he on a rocky ledge on the coast or aboard a ship?

Most probably he was on a ship since there is no surf in the foreground,

[a]A. A. Rylov, (1870–1939), Russian Soviet painter of epic romantic manner.

the distribution of waves is symmetrical and not deformed by the neighboring shore.

Let us try to estimate the velocity of the wind filling the sails of the ship floating at a distance. We are not the first interested by the question of evaluation of the speed of wind on the basis of the height of waves and other natural evidence. It was already in 1806 that Sir Francis Beaufort[b] introduced his approximate twelve-step scale. He related the force of the wind to the effects it had on land objects and to sea choppiness (see Table 4.1 on pp. 34–35). This scale has been approved by the International Meteorological Committee and is still in use.

Turning back to the picture we see that the sea is rather quiet. There are only few white horses on the surface. According to the Beaufort scale this corresponds to a gentle breeze with the velocity of about 10 *mph*.

One may judge velocity of wind not only by means of the Beaufort scale but from the brightness contrast between the sea and the sky. Usually the horizon in the open sea looks like a clear-cut border. The brightness of the sea becomes equal to that of the sky only at dead calm. In this case the contrast disappears and the sea becomes indiscernible from the sky. This happens rarely because the calm must be almost absolute, Beaufort number being zero. Even the slightest wind disturbs the sea surface. The coefficient of reflection from the oblique elements of the surface is no longer equal to unity and the contrast appears between the sea and sky. It may be studied experimentally. The dependence of the sea sky contrast on the wind force was measured during an expedition of the Russian research ship "Dmitrii Mendeleev". The results of measurements are depicted in Figure 4.1 by crosses, while the solid line represents the theoretical relation found by A. V. Byalko and V. N. Pelevin.

By the way, why are white watercaps so much unlike the blue-green sea water?

The color of the sea is defined by many factors. Among the most important are the position of the sun, the color of the sky, the form of the sea surface, and the depth. In shallow waters presence of sea-weeds and pollution by solid particles are relevant. All these factors affect reflection of light from the surface, its submarine scattering, and absorption. This makes unambiguous prediction of sea color impossible. Nevertheless some

---

[b]F. Beaufort, (1774–1857), English admiral, hydrographer, and cartographer, head of the English hydrographic service.

Fig. 4.1: Dependence of brightness contrast between the sea and the sky on wind speed.

details can be understood. For example, we can explain why the foreground waves that were closer to the painter were much darker in hue than others and why the sea became lighter near the horizon.

The extent of reflection of a light wave from the interface of two media with different optical densities is determined by the angle of incidence and the relative refraction index of the media. It is quantitatively characterized by the coefficient of reflection that is equal to the ratio of intensities of the incident and reflected rays[c]. The coefficient of reflection depends on incidence. In order to check this you may watch reflection of light from a polished table. The transparent varnish will act as the more dense optical medium. You will see that tangent rays are almost completely reflected, but the smaller the angle of incidence becomes the more light penetrates the optically dense medium and the less is reflected from the boundary. The coefficient of reflection falls with the decrease of the incidence angle.

Let us consider the schematic image of a wave in Figure 1 (refer to the color plate at the front). It is obvious that the incidence angles α_1 and α_2 of the rays coming to the observer from the front and back sides of a wave are different, $\alpha_2 > \alpha_1$. Hence more light comes to the eyes of the observer from the distant areas and the front faces of waves look darker than the plane sea far away. Certainly, in a troubled sea the angle α varies. However from far away the angular size of the darker crests decreases rapidly even though the angle α_2 remains bigger than α_1. Near the horizon the observer sees not solitary waves but the averaged pattern, the troughs between waves are

[c]Intensity of light is the averaged-over-time value of the light flux through a unit area perpendicular to the direction of the light propagation.

hidden and gradually the darker sides of the waves disappear. Because of this the sea near the horizon looks lighter than the sea in the foreground of the picture.

Now we can explain why watercaps are white. The seething water in the cap swarms with endlessly moving, deforming and bursting bubbles. Reflection angles vary from point to point and with time. As a result sunrays are almost fully reflected by the froth and watercaps appear white[d].

The color of the sea is greatly affected by the color of the sky. We have already said that the first is practically unpredictable. However the second can be understood from physical principles. Obviously the color of the sky is determined by the scattering of solar rays in the atmosphere of the Earth. The spectrum of sunlight is continuous and contains all wavelengths. Why does the scattering make the sky blue whereas the Sun seems yellow? We shall explain that with the help of the Rayleigh law of light scattering.

In 1898 the English physicist Lord Rayleigh[e] developed the theory of scattering of light by particles much smaller than the wavelength of the light. He found the law that states that *the intensity of the scattered light varies inversely with the fourth power of the wavelength.* In order to explain the color of the sky Rayleigh applied his law to the scattering of sunlight in the atmosphere. (For this reason the statement above is sometimes called the "law of the blue sky".)

Let us try to understand the meaning of the Rayleigh law qualitatively. Remember that light consists of electromagnetic waves. Molecules are built of charged particles, i.e. of electrons and nuclei. In the field of an electromagnetic wave these particles start moving. Displacement of bound electrons in the atom may be considered as harmonical: $x(t) = A_0 \sin \omega t$, where A_0 is the amplitude of oscillations and ω is the wave frequency. The acceleration of the particles is $a = -\omega^2 A_0 \sin \omega t$. Accelerated charged particles become sources of electromagnetic radiation themselves and emit the so-called secondary waves. The amplitude of the secondary wave is proportional to the acceleration of the emitting particle. (As you know uniformly moving charged particles make electric current but do not generate electromagnetic waves.) Therefore the intensity of the secondary emission is

---

[d]In the froth of watercaps blue rays coming from the sky are mixed with yellowish sunlight giving the white color.

[e]J. W. S. Rayleigh, (1842–1919), English physicist, chairman of the London Royal Society; Nobel Prize 1904 in physics.

proportional to the square of the acceleration of electrons in the field of the primary wave (one may neglect the motion of heavy nuclei) and therefore to the fourth power of the frequency, ($l \propto a^2 \propto (x_t'')^2 \propto \omega^4$).

Now return to the sky. The ratio of the wavelength of blue to that of red is $650\,nm/450\,nm = 1.44$ ($1\,nm$ (nanometer) $= 10^{-9}\,m$). The fourth power of this number is 4.3. Thus according to the Rayleigh law the intensity of the blue light scattered by the atmosphere exceeds that of the red four times. As a result, the air layer ten miles thick acquires a blue tint. On the contrary, the blue component of the sunlight that reaches us through the atmospheric "filter" is seriously depleted. Hence the sun rays penetrating the atmosphere get a yellowish tone. The coloration may enhance getting red and orange during sunset when the rays have to pass along a longer path through the air. (Obviously the colors change in the reverse order when the sun rises.)

It is interesting that despite the claims of the Rayleigh law, the wavelength of the scattered light to be much bigger than scattering particles, the intensity of scattering does not depend on the particle size. At first Rayleigh believed that the color of the sky is due to the tiniest dust polluting the atmosphere. But then he decided that sunrays are scattered by molecules of gases that make up the air. Ten years later in 1908 the Polish theoretical physicist M. Smoluchowski[f] proposed the idea that scattering is effected by rather unexpected objects, namely by inhomogeneities of the density of particles. With the help of this hypothesis Smoluchowski managed to explain the long-known phenomenon of *critical opalescence* – that is the strong scattering of light in liquids and gases that occurs near a critical point. Finally in 1910 Albert Einstein[g] formulated the consistent quantitative theory of molecular scattering of light that leaned upon the ideas of Smoluchowski. In the case of gases, the intensity of scattered light exactly coincides with the earlier result by Rayleigh.

It seems that everything is in order now. But what is the origin of inhomogeneities in the air? Supposedly air must be in thermodynamical equilibrium. Gigantic inhomogeneities that make wind blow are incomparable with wavelengths of light and can not affect the scattering.

In order to clear up the origin of the inhomogeneities we must discuss

---

[f]M. von Smolan-Smoluchowski, (1872–1917), Polish physicist, classical works on fluctuation theory, and theory of Brownian motion.

[g]See footnote on page 5.

the concept of thermodynamical equilibrium in more detail. For simplicity let us consider a macroscopic amount of gas confined in a closed volume.

Physical systems consist of enormous numbers of particles. This makes statistical description the only possible approach. Statistical treatment means that instead of following separate molecules we calculate average values of physical characteristics of the whole system. It is not at all necessary that the corresponding value is the same for all molecules. The most probable distribution of molecules in the macroscopic gas volume would be the uniform one. But because of thermal motion there is a nonzero probability that the concentration of molecules in some part of our container will be enhanced (and as a result it will fall somewhere else). Theoretically it is even possible that all molecules will assemble in one half of the container leaving another half absolutely empty. However the probability of such an event is expressed by an extremely small number. So there is little hope realizing it even once in the 10^{10} years that is currently believed to be the lifetime of the Universe.

Small deviations of physical quantities from their averages are not only allowed but happen perpetually due to the thermal motion of molecules. These deviations are called *fluctuations* (in Latin *fluctus* means wave). Because of fluctuations the gas density may be greater here and less there and as a result the refraction index will vary from place to place.

Let us now return to the scattering of light. All the reasoning applied to the closed container will hold in the atmosphere. Light is scattered by inhomogeneities of the refraction index that come from density fluctuations. Moreover, air is the mixture of several gases. Distinctions in thermal motion of different molecules provide another source of inhomogeneities.

The typical scale of inhomogeneity of the refraction index (and of those of density) depends on temperature. The sunlight is mainly scattered in the atmospheric layers where inhomogeneities are much less than wavelengths of visible light but much greater than the molecules of gases that compose the air. This means that scattering is effected by inhomogeneities but not by molecules, as it was presumed by Rayleigh.

Nevertheless the sky is blue and not violet contrary to the prediction of the Rayleigh law. There are two reasons for this discrepancy. First, the spectrum of the Sun contains much fewer violet rays than blue ones. The second thing responsible for the seeming disagreement of the theory with practice is our "registering device", the ordinary human eye. It turns out that visual perception markedly depends on the wavelength of light. The experimental curve characterizing this dependence is plotted in Figure 2 (refer to color plate at the front). It is clear that our eye responds to violet rays much more weakly than to blue and green ones. This conceals the violet component of the scattered sunrays from the human eye.

Well, then why are the clouds that we see in the sky unmistakably white? Maybe they are composed of particles that violate the Rayleigh law and our conclusions are no longer true?

Clouds consist either of water drops or of ice crystals that in spite of being microscopic are much larger than the visible wavelengths. The Rayleigh law is not valid for these particles and the intensity of the scattered light is nearly the same for all wavelengths. This makes clouds like those painted in the picture.

Now turn your attention to the form of the clouds. The tops of the clouds in the picture are drawn fluffy and wreathing (these are the so-called heap clouds) but the lower surfaces are plainly outlined. For what reason? It is known that heap clouds (in distinction to sheet ones) are formed by uprising convective flows of humid air. The temperature of air decreases with height above the sea (as well as above the land). As long as the altitude is much less than the Earth's radius and the distance to the nearest shore surfaces of constant temperature (the isotherms) are next to horizontal plains parallel to the sea surface, the temperature drop is sufficiently fast near the sea level, being about $1°\,C$ per hundred meters (this makes about $= 1.6°\,F$ per $100\,yd$). (Generally speaking the dependence of the air temperature on the altitude is far from linear but for elevations less than several miles these numbers are correct.)

Now, what happens to the upward flow of humid air? As soon as it reaches the altitude where the temperature of air corresponds to the dew point of the vapor, the water it carries begins condensing to tiny drops. The isotherm where this happens contours the bottom of the cloud. Irregularities of the surface where condensation takes place stay within some tens of meters whereas clouds stretch for hundreds and thousands of meters. Hence their lower bounds are almost flat. This is confirmed by the row of clouds in the distance hovering horizontally above the sea.

However the rise does not stop at the bottom of a cloud. The air continues the upward motion and cools rapidly. The remaining vapor suffers intensive condensation, first into droplets and then into little ice crystals. These crystals usually form the top of heap clouds. After having lost the vapor and cooling down, the air slows down and turns back. It flows sideways and around the cloud. Convective flows lead to the formation of the characteristic curls at the tops of heap clouds and the descending cold air keeps clouds apart. Thanks to that they do not merge into a heavy grey mass, being interspaced by blue intervals.

You see in the foreground a flight of white birds. Let us estimate the frequency of wing flaps of a medium-sized bird (say with the mass $m \approx 10\,kg$ and the area of wings $S \approx 1\,m^2$) when it flies without gliding. Let the average velocity of the wings be $\bar{v}$. Then in the time Δt the bird wings will render velocity $\bar{v}$ to the mass $\Delta m = \rho S \bar{v} \Delta t$ of the air (here ρ is the air density). The total momentum passed by the wings to the air will be $\Delta \bar{p} = \rho S \bar{v}^2 \Delta t$. In order to keep at the same height we must compensate the weight of the bird. This means that $\Delta \bar{p} = m g \Delta t$. We may conclude that

$$m g = \rho S \bar{v}^2,$$

and the mean velocity of the moving wings is $\bar{v} = \sqrt{m g / \rho S}$. This velocity can be related to the frequency ν of flaps and the length of the wings L in the usual way:

$$\bar{v} = \omega L = 2\pi \nu L.$$

Assuming that $L \sim \sqrt{S}$ we find:

$$\nu \approx \frac{1}{2\pi S} \sqrt{\frac{m g}{\rho}} \approx 1\,s^{-1}.$$

Thus according to our calculation the bird should flap the wings once per second and that looks quite a reasonable estimate by the order of magnitude.

It is interesting to discuss the outcome formula in more detail. Let us suppose that all birds have roughly the same body form regardless to the size and species. Then one may link the area of the wings to the mass of the bird by the relation $S \propto m^{2/3}$. Substituting this into the earlier found expression for the frequency of the wing flaps we obtain that

$$\nu \propto \frac{1}{m^{1/6}}.$$

From this we conclude that the frequency of flaps grows with the decrease in the mass of birds. This absolutely agrees with the common sense explanation. Certainly the assumption that all birds have wings of the same form is extremely rough since the wings of most big birds are relatively bigger than those of small ones. Nevertheless this only supports the trend.

Let us note that the same formula (3.4) could be derived using the dimensional approach (except the important factor 2π). It is clear that the frequency of wing flaps depends on the bird's weight, the area of the wings S, and the density ρ of the ambient air. Let us search for a relation between the four of them. Suppose that $\nu = \rho^\alpha S^\beta (m g)^\gamma$ with α, β, γ being unknown numbers. Comparison of the dimensions of the quantities in both sides of the relation gives $\alpha = -\gamma = -1/2$ and $\beta = -1$. From here follows that

$$\nu \sim \frac{1}{S} \sqrt{\frac{m g}{\rho}}.$$

Questions and answers that can be found *"out of the blue"* are far from being exhausted. Curious and observant readers will perhaps find in the picture other more instructive aspects. However why should we limit ourselves by the frame of the picture? There are plenty of interesting questions and problems in the everyday world all around us.

Can you explain why astronauts say that from outer space the Earth looks blue?

Table 4.1: BEAUFORT WIND SCALE

Beaufort Force Number	State of Air	Wind velocity (knots)	(mph)	Description of sea surface
0	calm	0–1	< 1.15	sea like a mirror
1	light airs	1–3	1.2–3.5	ripples with appearance of scales are formed, without foam crests
2	slight breeze	4–6	4.6–6.9	small wavelets still short but more pronounced; crests have a glassy appearance but do not break
3	gentle breeze	7–10	8.0–12	large wavelets; crests begin to break; foam of glassy appearance; perhaps scattered white horses
4	moderate breeze	11–16	13–18	small waves becoming longer; fairly frequent white horses
5	fresh breeze	17–21	20–24	moderate waves taking a more pronounced long form; many white horses are formed; chance of some spray
6	strong breeze	22–27	25–31	large waves begin to form; the white foam crests are more extensive everywhere; probably some spray
7	moderate gale	28–33	32–38	sea heaps up and white foam from breaking waves begins to be blown in streaks along the direction of the wind; spindrift begins to be seen
8	fresh gale	34–40	39–46	moderately high waves of greater length; edges of crests break into spindrift; foam is blown in well-marked streaks along the direction of the wind

Table 4.1: BEAUFORT WIND SCALE *(continued)*

Beaufort Force Number	State of Air	Wind velocity		Description of sea surface
		(knots)	(mph)	
9	strong gale	41–47	47–54	high waves; dense streaks of foam along the direction of the wind; sea begins to roll; spray affects visibility
10	whole gale	48–55	55–63	very high waves with long overhanging crests; resulting foam in great patches is blown in dense white streaks along the direction of the wind; on the whole the surface of the sea takes on a white appearance; rolling of the sea becomes heavy; visibility affected
11	storm	56–65	64–75	exceptionally high waves; small- and medium-sized ships might be for a long time lost to view behind the waves; sea is covered with long white patches of foam; everywhere the edges of the wave crests are blown into foam; visibility affected
> 12	hurricane	> 65	> 75	the air is filled with foam and spray; sea is completely white with driving spray; visibility very seriously affected

Chapter 5

The moon-glades

Reflections of various light sources from the surface of water often look like long shimmering lanes leading from the source to our eye. Remember the setting sun reflected by sea or street lights along a night river quay. The glittering of the moon adorns the sea or lake with a wide stripe of light.

All this happens because every wavelet on the surface gives a separate image of the source. Let us try to understand why reflections from thousands of illuminated ripples make a glade, that is, an oblong figure directed from the light source to the observer.

As you already know, wavelets are formed at Beaufort numbers between 1 and 3. At weaker winds water is calm and the surface reflects like a plain mirror. Stronger winds bring on foam and white horses and the contour of the glade becomes vague. One may visualize ripples as scores of wavelets running chaotically in all directions[a]. Slopes of their surfaces do not exceed some limiting value α which depends on the wind and can reach $20°$–$30°$.

Now let us simplify the problem a bit and try to find the proportions of the glade. Suppose that everywhere on the surface there is plenty of tiny mirror-like ripples looking in all directions. The slopes of wavelets range from 0 to α (this follows from the smoothness of the surface). For simplicity we shall assume that both the light source and the observer are at the same height h above the water, Figure 5.1.

A small horizontal mirror will cast the beam to the eye of the observer only if its distance to the source and to the observer are equal. This means it must be at the point M. On the other hand a mirror tilted at the angle

---

[a]Remember the Raileigh independence principle —A. A.

Fig. 5.1: The length of glade is determined by the steepest slopes turned toward and away from the observer. The width of glade is defined by reflections from slopes tilted sidewards.

α towards the observer must be moved away from him to the point N. Conversely a mirror tilted away from the observer should be transferred to the closer point N'.

The tilted mirrors imitate the extreme positions from where waves still reflect light to our eyes. Therefore the distance from N to N' determines the length of the light-glade. Everywhere in between one can find waves with the right slope which reflect rays to the observer.

Now let us consider angles between the light rays. One can notice from Figure 5.1 that $\beta + \alpha = \gamma + \delta$, $\beta - \alpha = \varepsilon = \delta$ and $\gamma = \alpha + \beta - (\beta - \alpha) = 2\alpha$. Hence we may conclude that the angular size of the longer axis of the light spot simply equals the angle between the steepest slopes of the ripples. It is no problem to calculate the length $N\,N'$ of the glade.

The shorter axis of the reflecting light spot is easily calculated in a similar manner. Let us shift the mirror away from the central point M in the direction transverse to $N\,N'$. In order to reflect beams to the eye the mirror must be turned around the axis parallel to $N\,N'$, Figure 5.1. Having in mind that the maximal tilt of the mirror remains α, we find that the width of the light strip is $P\,P' = 2h\tan\alpha$ and hence the shorter axis is seen at the angle $\beta = 2\arctan\frac{h\tan\alpha}{\sqrt{l^2+h^2}} \approx \frac{2h\tan\alpha}{\sqrt{l^2+h^2}}$.

The ratio of the apparent sizes of the two axes is $\beta\,/\,2\alpha$. If the spot is not too large and the angle α is small then $\beta\,/\,2\alpha \approx \frac{h}{\sqrt{l^2+h^2}} = \sin\omega$ where ω

is the incidence angle at the point M[b]. As the angle ω becomes smaller the spot grows longer. If the look glances along the surface the spot appears infinitely long and narrow.

When we watch moon-glades on the sea, the angle ω is most often small and the glade stretches to the horizon, Figure 5.2. Of course, here our formulae are not literally applicable. Nevertheless they help to explain qualitatively the origin of glades and understand the effects of wind speed and the moon's altitude above the horizon on glade width: increasing α and h makes the glade wider.

Fig. 5.2: The wind speeds are (from left to right): $12\,m/s$; $12\,m/s$; $5\,m/s$; $2\,m/s$. The altitudes of the sun above horizon are: $30°$; $20°$; $13°$; $7°$.

What is the value of the angle ω if the angular altitude of the moon is β?

---

[b]To be specific M is the point where the reflection from the calm waters would be. —A. A.

Chapter 6

The Foucault pendulum and
the Baer law

...There was something, however, in the appearance of
this machine which caused me to regard it more atten-
tively.

Edgar Allan Poe, *"The Pit and the Pendulum."*

Those lucky enough to have been to St. Petersburg will remember the
famous pendulum in St. Isaac's Cathedral[a]. Others might have heard
about it (Figure 6.1). The swings of the pendulum are accompanied by the
slow rotation of the plane of oscillations. This observation was first made
in 1851 by the French scientist J. Foucault[b]. The experiment was carried
out in the spacious hall of the Pantheon in Paris, the ball of the pendulum
had the mass of $28\,kg$, $(68\,lb)$ and the string was 67 meters $(73\,yd)$ long.
Since then this sort of pendulum are named after Foucault. How could one
explain its motion?

You know from the textbooks that if Newton's[c] laws were true on the
Earth then the pendulum would keep to the plane of oscillations. This
means that in the reference frame rigidly bound with the Earth the laws
of Newton must be "corrected". In order to do this one has to introduce
special forces called *inertia forces*.

---

[a]St. Petersburg, the city and seaport on the Gulf of Finland of the Baltic Sea was during
1712–1917 the capital of Russian Empire. In 1917, after revolution, it was renamed
Leningrad but now bears the original name.

[b]J. B. L. Foucault, (1819–1868), French physicist, foreign member of Russian Academy
of Sciences.

[c]Isaac Newton, (1642–1727), English philosopher and mathematician; formulator of the
laws of classical mechanics and gravity.

Fig. 6.1: In March 1931 the Foucault pendulum was
first presented in St. Isaac's Cathedral occupied
at that time by the Leningrad Antireligious
Museum.

6.1   Inertia forces in the rotating reference frame

Inertia forces must be introduced in any reference frame that accelerates
with respect to the Sun (or, to be precise, to the so-called stationary stars).
These are called non-inertial reference frames in distinction to inertial ones
that move uniformly with respect to the Sun and stationary stars.

Strictly speaking, the Earth does not present an inertial frame of reference because it orbits the Sun and revolves itself. Usually one may neglect accelerations arising from these motions and apply Newton's laws. However, this is not the case with a Foucault pendulum. Precession of the oscillation plane is explained by the action of the special force of inertia called the Coriolis force[d]. Let us examine it.

Here is a simple example of a rotating reference frame where inertia forces reveal themselves clearly.

Imagine a man riding a merry-go-round in Gorky Park[e]. Let the radius of the circle be r and the angular velocity of rotation ω. Suppose that the man tries to jump from his seat to the one in front of him, Figure 6.2, moving with velocity v_0 with respect to the platform.

Fig. 6.2: Inertia forces in revolving reference frame.

⚠️ **Warning! The experiment is purely imaginary being strictly forbidden by safety regulations.**

First let us consider the motion of our hero in a stationary reference frame. Obviously the motion is circular with the linear velocity v that adds up the linear velocity ωr of the merry-go-round and his relative velocity:

$$v = \omega r + v_0.$$

---

[d]G. G. Coriolis, (1792–1843), French civil engineer.

[e]See the novel by M. Cruz-Smith for further reference.

The centripetal acceleration is defined by the common formula,

$$a_{\text{cp}} = \frac{v^2}{r} = \frac{v_0^2}{r} + \omega^2 r + 2 v_0 \omega.$$

According to Newton's second law, the acceleration is due to the horizontal component of the force exerted onto the man by the rotating platform, the seat, handles etc.,

$$m\, a_{\text{cp}} = Q.$$

Now consider the motion in the reference frame bound to the merry-go-round. Here the linear velocity is v_0 and the centripetal acceleration is $a'_{\text{cp}} = \frac{v_0^2}{r}$. With the help of the two previous equalities we may write:

$$m\, a_{\text{cp}} = \frac{m\, v_0^2}{r} = Q - m\,\omega^2 r - 2m\, v_0\, \omega.$$

In order to apply the second Newton law in the revolving frame of reference we must introduce the force of inertia:

$$F_{\text{in}} = -(m\,\omega^2 r + 2m\, v_0\,\omega) = -(F_{\text{cf}} + F_{\text{Cor}}),$$

where the minus sign indicates that it is directed away from the axis. In the non-inertial frame the equation of motion will be:

$$m\, a'_{\text{cp}} = Q + F_{\text{in}} = Q - (F_{\text{cf}} + F_{\text{Cor}}).$$

It seems that the inertia force throws you from the center of the merry-go-round. However the word "seems" is not a slip. No new interactions between the bodies appear in the rotating reference frame. The only real forces acting on the man are the same reactions of the seat and bars. Their net horizontal component Q is directed towards the center. In the stationary reference frame the force Q resulted in the centripetal acceleration a_{cp}. In the rotating frame due to kinematical reasons the acceleration changed to the smaller value a'_{cp}. In order to restore the balance between the two sides of the equation we had to introduce the force of inertia.

In our case the force F_{in} consists of the two addends. The first is the centrifugal force F_{cf} that increases with the frequency of rotation and with the distance from the center. The second is the Coriolis force F_{Cor} named after the person who first calculated it. This force has to be introduced only when the body moves relative to the rotating frame. It depends not on the position of the body but on its velocity and the angular velocity of rotation.

If the body in the rotating frame moves not along a circle but radially, Figure 6.2, then, just the same, one must introduce the Coriolis force. Now it is perpendicular to the radius unlike the previous case. One of the basic features of the Coriolis force is that it is always perpendicular both to the axis of rotation and to the direction of motion. It may look strange but in the revolving frame inertia forces not only push a body away from the center but tend to swerve it astray.

We must emphasize that the Coriolis force, like all other inertia forces, is kinematical in origin and can not be related to any physical objects[f]. Here is an explicit example.

Imagine a cannon set at the North pole and pointed along a meridian (the pole is chosen for simplicity). Let the target lie on the same meridian. Is it possible that the projectile hits the target? From the point of view of an external observer which uses the inertial frame bound to the Sun, the answer is obvious: the trajectory of the projectile lies in the initial meridional plane whereas the aim revolves with the Earth. Thus the projectile will never get the target (unless a whole number of days will elapse). But how could one explain the fact in the reference frame bound to the Earth? What causes the projectile to stray from the initial vertical plane? In order to restore consistency one has to introduce the Coriolis force that is perpendicular to the rotation axis and to the velocity of a body. This force pulls the projectile away from the meridional plane and it misses the target.

Now let us return to the precession of the oscillation plane of the Foucault pendulum from which we have started. It is caused by a quite similar reason. Suppose again that the pendulum is situated at the pole. Then for a stationary observer the oscillation plane is at rest and it is the Earth that rotates. A denizen of the North pole will see the opposite. For him the meridional plane looks fixed whereas the oscillation plane of the pendulum performs a full revolution every 24 hours. The only way to explain this is with the help of the Coriolis force. Unfortunately, in general, the picture is not so transparent as at the pole[g].

---

[f]Even though the force of inertia is not produced by any real bodies, observers feel it as a real force, akin to gravity. Remember the centrifugal force in the turning car.

[g]Oscillation plane of a Foucault pendulum located elsewhere turns $2\pi \sin \alpha$ radians per day, where α is the latitude of the place.

6.2   Interesting consequences

The Coriolis force which appears due to the Earth's rotation leads to a number of important effects. But before discussing those let us establish the direction of the Coriolis force. We have said that it is always perpendicular to the rotation axis and to the velocity of motion. However this leaves two possibilities depicted in Figure 6.3. Remember that an analogous situation emerges when defining the direction of the Lorentz[h] force exerted onto a moving charge by a magnetic field. You may remember from textbooks that it is perpendicular to the velocity of the charge and to the magnetic induction. Still in order to define it unambiguously one has to resort to the *left-hand rule*.

Fig. 6.3: Two options for the direction of
Coriolis force. By convention the direction is
fixed by the left-hand rule.

The direction of the Coriolis force can be determined by means of a similar rule elucidated in Figure 6.3, *a*. First of all we must assign a direction to the axis. By convention looking in this direction one sees the clockwise rotation[i]. Now let us pose the left hand with the four fingers pointing in the direction of the velocity so that the axis pierce through the palm. The thumb positioned at a right angle will show the direction of the Coriolis force.

The alternatives in defining the directions of Coriolis and Lorentz forces correspond to the two kinds of symmetry encountered in nature, the left

[h]H. L. Lorentz, (1853–1928), the Dutch physicist; Nobel Prize 1902 in physics.

[i]The *gimlet* rule states that this is where a gimlet rigidly attached to the frame would move.

and right symmetries. In order to classify the symmetry one has to use "standards" such as hand, gimlet, cork-screw, etc. Certainly nature does not care about your hand or gimlet. These are simply tools that help to fix the direction of the force.

This completes the discussion of the Coriolis force for the case when the velocity of a body in the revolving frame is perpendicular to the axis. The magnitude of the force is $2m\,\omega\,v_0$ and the direction is defined by the left-hand rule. But what happens in the general case?

It turns out that if the velocity v_0 makes an arbitrary angle with the rotation axis, Figure 6.3, *b*, then only the projection of v_0 onto the plane perpendicular to the axis is important. The value of the Coriolis force is given by the following formula:

$$F_{\text{Cor}} = 2m\,\omega\,v_\perp = 2m\,\omega\,v_0\cos\phi.$$

The direction of the force is determined by the same left-hand rule although now the fingers must be parallel not to the velocity, but to its projection onto the plane perpendicular to the axis, Figure 6.3, *b*.

Now we have learned everything about Coriolis force: both how to calculate the value and to define the direction. Armed with this knowledge we may explicate a number of interesting effects.

It is well known that trade winds which blow from the tropics to the equator are always deflected westward. This effect is explained in Figure 6.4. First let us consider the Northern hemisphere where trade winds blow from north to south. Position the left hand above a globe, the palm down. The axis enters the palm being perpendicular to the fingers. The Coriolis force is perpendicular to the page being leveled at you, that is to the west. Trade winds of the Southern hemisphere in their own turn blow from the tropic of Capricorn north to the equator. However, neither the direction of rotation nor the projection of wind velocity onto the equatorial plane change. Therefore the direction of the Coriolis force does not change either and in both cases the winds are diverted to the west.

Figure 6.5 illustrates Baer's law[j]. The right banks of rivers in the Northern hemisphere are steeper and more undermined than the left ones (and vice versa in the Southern hemisphere). The reason is again the Coriolis force that pushes flowing water to the right. Because of friction the surface

---

[j]K. E. von Baer, (1792–1876), Estonian zoologist and pioneer embryologist who was among the founders of the Russian Geographical Society.

Fig. 6.4: Coriolis force deflects trade winds to
the west.

velocity of a stream is bigger than that at the bottom; hence the Coriolis
force is also bigger. This gives rise to the circulation of water shown in
Figure 6.6 by arrows. The soil of the right bank is washed away and settles
at the left side. This strongly resembles the erosion of the bank at river
turns which was described in the chapter dedicated to meanders.

Fig. 6.5: Coriolis
force drives water
flows to the right in
the Northern
hemisphere and to the
left in the Southern.

Coriolis force always leads to eastward deviation of falling bodies.
(Tackle this yourself.) In 1833 the German physicist Ferdinand Reich carried out precision experiments in the Freiburg mine. He found that the

Fig. 6.6: The right banks of rivers in the Northern hemisphere are steeper and more undermined than the left ones.

Right bank

Left bank

Northern hemisphere

average (over 106 measurements) deflection of bodies which were dropped from the height of 158 m (178 yd) was 28.3 mm (1.11 in). This was one of the first experimental proofs of the Coriolis theory.

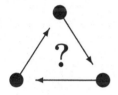

Try to estimate the difference of water levels at the right and left sides of the Volga river.

Does the Baer law apply to rivers streaming along parallels or the equator? What changes if a river crosses the equator, like the Congo?

Chapter 7

The moon-brake

Time and tide wait for no man.

A proverb.

Long ago people identified the Moon as the reason for tides. The Moon attracts the water of the world ocean and forms a water "hump" in the ocean. The hump keeps its place on the moon side as the Earth rotates about the axis. When the high water advances to a coast the tide rises and when it retreats the ebb starts. The theory looks rather natural but it leads to a contradiction. This would mean that tides must be a daily event but instead, they happen every twelve hours.

The first explanation was given by the Newton's theory of tides which appeared shortly after the discovery of the law of gravity. We shall study this question using the idea of inertia forces. According to the previous chapter one may apply Newton's laws of mechanics in a revolving reference frame after adding to interactions between physical bodies the forces of inertia.

The Earth rotates around its axis, around the Sun and around... the Moon. Usually one forgets the latter, but it is this rotation that makes possible the theory of tides. Imagine that two balls, one light and one heavy, linked by a string are placed on a smooth surface, Figure 7.1.

Rotations of the tied balls are interrelated. Each one follows a circle of its own radius but the common center of the two is at the center of mass of the system. Of course the bigger ball traces the smaller circle but it moves! Similarly, the Moon and the Earth are attracted to each other, according to

51

Fig. 7.1: Two linked
balls revolve around
the common center of
mass.

the law of gravity, and are orbiting their common spatial center of inertia
C, Figure 7.2. Because of the big mass of the Earth, this point lies inside
the globe being shifted with respect to the center O. The angular velocities
of rotation around C of both the planet and the satellite are evidently the
same.

Fig. 7.2: The Earth and the Moon are orbiting the
common center of inertia.

Consider now the revolving reference frame where both the Earth and
the Moon stay at rest. Inasmuch as the reference frame is noninertial every
mass element experiences not only the force of gravity but a centrifugal force
as well. The farther from the center C the stronger this force becomes.

Let us imagine for simplicity that water is evenly spread over the entire
surface of the globe. May this be in equilibrium? Obviously not. Gravitational attraction to the Moon and centrifugal forces will destabilize the
state. On the moon side, the two forces are directed away from the Earth's

center and give rise to the water hump A, Figure 7.2. But the situation on the remote side is quite alike. As we step away from the common center of mass, the centrifugal force increases whereas the attraction to the Moon falls down. The resultant force is again directed away from the center of the Earth and is the origin of the second hump B. The equilibrium configuration is represented in Figure 7.2.

Of course this explanation of tides is much simplified. It does not take into account the nonuniform distribution of water on the Earth, effects of attraction to the Sun and many other factors that may essentially influence the picture. Still the theory answers the chief question. Once the humps do not move (with respect to the rotating frame of reference) but the planet revolves around its own axis the tides must occur twice a day.

Now we should explain the principle of the moon-brake. It turns out that the humps actually lie not on the line connecting the centers of the Earth and the Moon (as it was shown for simplicity in Figure 7.2) but are a little displaced, Figure 7.3. The reason is that because of friction, the ocean rotates together with the Earth. Therefore the mass of water in the humps is continuously renewed. However the deformation is always in retard with respect to the force that brings it on. (The force gives rise to an acceleration but it takes time for particles to gain speed and reach the place.) Thus the top of the hump that is the point of the highest tide is not at the point of the strongest attraction to the Moon which lies on the line connecting the centers. The hump is formed with a lag and it is shifted in the direction of the daily rotation of the Earth. According to Figure 7.3 this implies that the force of gravitational attraction to the Moon does not pass through the center of the Earth and brings about a torque that slows down the gyration. The duration of the revolution is enhanced daily! This phenomenon was first recognized by the wonderful English physicist Lord Kelvin[a].

The "moon-brake" has operated flawlessly for many millions of years and has the capability to notably change the length of the day. Scientists discovered in corals that have lived in the ocean about 400 million years ago, structures called "diurnal" and "annual" rings. When the diurnal rings were counted it turned out that there were 395 of them per year! The length of a year, that is the period of the circumvolution of the Earth around the

---

[a]W. Thomson, (1824–1907), 1st Baron Kelvin since 1892; English physicist and mathematician, chairman of the London Royal Society.

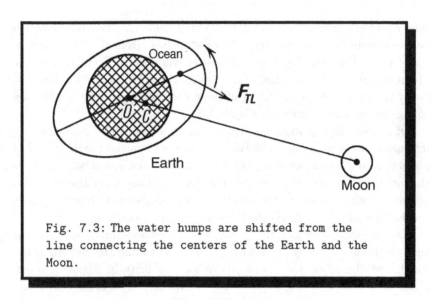

Fig. 7.3: The water humps are shifted from the
line connecting the centers of the Earth and the
Moon.

Sun, has probably not changed since then. Hence in those times the day
lasted only 22 hours!

Now the moon-brake keeps working on making days and nights longer.
At the end of the story, the period of the Earth's daily rotation will get
equal to that of the Moon orbiting and the impeding will cease. The Earth
will forever remain turned to the Moon by the same side, as presently the
Moon is. The increase of the day will effect the climate. The extended day
on the sunny side of the globe will be opposed by the prolonged night on
the rear. Cold air from the night side will rush to the warm hemisphere.
Winds and dust-storms will break out... But this prospect is so far off that
mankind will definitely discover how to prevent these calamities.

*What is the effect of the moon-brake on the du-
ration of the lunar month (that is the period of
orbiting of the Earth by the Moon)?*

PART 2

Saturday night physics

We've gotten so accustomed to our circumstances that we don't even notice the many wonders and think about the actual causes behind them. Yet at closer inspection, one discovers a great many grounds for contemplation.

"When tossing pebbles into water, focus on the circles they produce, otherwise your tossing will be a mere time-frittering," — wrote the great Koz'ma Prutkov[a].

We shall try to convince you that even the most improbable phenomena in the world around us can be explained by ordinary physics.

---

[a]Koz'ma Prutkov was the great Russian writer, poet and philosopher of the nineteenth century. His selected works along with those by Alexander Pushkin and Mikhail Lomonosov may be found on the desk of any Russian scholar.

Chapter 8

Why the violin sings

The violin had no color, but sound it had.

N. Panchenko, *"The poem about a violin."*

Whenever an object moves through a medium, there always arise resistance forces trying to slow the motion. When a body slides mechanically along a rigid surface it will be the force of dry friction; in a liquid or in a gas it will be liquid friction (viscosity) and in the air it will be the aerodynamic resistance.

The interaction between a body and the surrounding medium is a rather complicated process leading usually to the work-to-heat conversion of mechanical energy of the body. However, the reverse situation, when the medium is in fact supplying the body with energy, is possible too. And this usually leads to some sort of oscillation. For example, the dry friction force between a wardrobe being moved and the floor will brake it, slowing the motion. The same friction between the bow and string of a violin will make the string reverberate. As we will see later, the cause of vibration in the latter case is the dropping dependence of friction on the velocity of the motion. The vibration indeed occurs when friction decreases with augmenting velocity.

Let's illustrate the generation of mechanical oscillations using as an example a concerto violin. The sound of a violin is caused by the moving bow, right? It's impossible, of course, to explain here all the complicated phenomena involved in formation of a particular musical tone, yet let us try to understand, in principle, why the string starts vibrating when the bow is being smoothly pulled against it.

The friction force between the bow and string is the dry friction. We can easily distinguish two different kinds of friction – friction of rest and sliding friction. The first acts between the touching surfaces when contacting bodies are at rest with respect to each other; the second – when one body is actually sliding along the surface of the other.

As it's known, in the former case (no sliding), friction will balance an external force (being equal in magnitude and opposite in direction) up to a certain maximal value, called F_{fr}^0.

The sliding friction, in its turn, depends on the material and condition of the contacting surfaces, as well as on the relative velocity of the bodies. It is the latter circumstance that we will discuss in more detail. The character of relation between sliding friction and velocity varies for different bodies: as velocity rises the sliding friction at first often drops down and then begins to go up too. Such dependence of the dry friction force on velocity is illustrated by the graph in Figure 8.1. The friction force between the hair of the bow and the string behaves in this way too. When v, the relative velocity of the bow and string, is zero the friction between them doesn't exceed F_{fr}^0. Then, for the descending part of the curve, $0 < v < v_0$, any slight increase of the relative speed, say, by Δv, leads to the corresponding decline of friction force and, vice versa, when velocity is going down the change of force will be positive (see Figure 8.1). And, as we are going to see in a minute, it's exactly due to this not so evident at first glance feature that the energy of the string can grow at the expense of mechanical work done by the force of dry friction.

Fig. 8.1: Typical dependence of dry friction on relative velocity.

When the bow starts its motion, the string is getting drawn along with it, and the friction is compensated by the tension of the string, Figure 8.2. The resultant of the tension forces is proportional to the deviation x of the string from the equilibrium:

$$F = 2T_0 \sin \alpha \approx \frac{4T_0}{l} x,$$

where l is the length of the string and T_0 is the tension force, which for small stretches, x, may be taken to be constant. Thus, when the string is being pulled along with the bow, the force F is growing until it reaches the maximal value of friction, F_{fr}^0. Then the string begins sliding against the bow.

Fig. 8.2: When string follows the bow without sliding the friction of rest compensates the resultant of the two tension forces.

For the sake of simplicity, let's assume that at the beginning of slipping the friction force drops abruptly from the maximum rest value F_{fr}^0 down to the relatively weak force of sliding friction. In other words, we can approximately consider the slipping of the string as an almost free motion.

At the exact instant when the string takes off from clinging to the bow to sliding, its velocity is equal to the that of the bow, and, therefore, it keeps moving in the same direction. Yet now, the net tension force, not compensated by anything, will start slowing the motion of the string down. Consequently, at a certain moment the velocity will drop to zero, the string will stop and then it will reverse its motion and go back against the bow. Further, after the maximal swing to the other side, the string will again have to start moving in the same direction with the bow. Throughout this time the bow continues to move with the same constant velocity u and at some point the speeds of string and bow will match both in their magnitude and, this time, in the direction too. The slipping between the string and the bow will disappear and, again, the friction will balance the string tension. Now, as the string approaches the neutral position, the tension force subsides causing the corresponding waning of friction that counteracts it. And then, after the string passes the equilibrium position, everything happens again.

The ensuing graph for the string deviation versus time is shown in Figure 8.3, *a*. The periodic motion of the string is composed of two different parts for each period. That is, for $0 < t < t_1$ the string follows the bow with constant speed u and the deviation x is linearly proportional to the elapsed time ($\tan \alpha = u$). At t_1 the "take-off" occurs and for the interval $t_1 < t < t_2$ the dependence of x on time becomes a sinusoid. At the instant t_2, when the tangent to the sinusoid has the same slope α as the starting linear piece of the curve (and, hence, the velocities of the string and the bow are equal), the string is captured by the bow again.

Fig. 8.3: Dependencies of string displacement on time: *a*--- in absence of sliding friction; *b*--- with nonzero sliding friction.

Figure 8.3, *a* illustrates an ideal case when there is no sliding friction between the bow and the string and, consequently, there is no energy loss as the string moves freely. The total work performed by friction forces (in the intervals without sliding) during the complete cycle of oscillation equals zero because for negative x the friction force acts against the motion and the mechanical work is negative, whereas for $x > 0$ the work is the same in magnitude yet positive in sign.

Now let's try to figure out what happens if the sliding friction force is not negligibly small any more. Then it should cause energy loss. The graph for the string motion with sliding is presented in Figure 8.3, *b*. For positive values of x the curve is actually steeper than for negative ones. Hence, now the clinging of string to the bow happens at a smaller negative deviation ($-x_2$ in the picture) than the positive x_1 at which the string first starts slipping against the bow: $x_2 < x_1$. This results in the positive mechanical work done by the friction in the intervals when the bow and the string go along together:

$$A = \frac{k(x_1^2 - x_2^2)}{2}.$$

Here $k = \frac{4T}{l}$ is the coefficient of proportionality which relates the value of friction of rest that pulls the string away from the equilibrium and the string swing[a].

This positive fraction of the total work compensates for energy losses due to the sliding friction and makes the string oscillate without damping.

Generally speaking, to replenish the energy, it's not at all necessary for the string to cling to the bow. It's enough for their relative velocity v to stay within the falling part of the sliding friction dependence on relative speed (see Figure 8.1). Now let's take a closer look at the vibration of the string in this case.

Suppose the bow is being pulled with constant speed u and the string is driven away from its neutral equilibrium position by x_0 so that the net tension force $F(x_0)$ is again compensated by the sliding friction force $F_{\text{fr}}(u)$. If, by chance, the string deviates in the direction the bow moves, their relative velocity shall decrease. This will cause friction to rise. (Note that we are on the dropping part of the $F_{\text{fr}}(u)$ curve!) This, in turn, makes the string stretch even more. Of course, at some point the elastic force will exceed the friction (remember that the vector sum of tensions is directly proportional to the deviation of string from the neutral position, whereas the friction is limited by F_{fr}^0). There the string will start slowing down, then it will reverse its motion and go in the opposite direction. On the way back the string will pass the point of equilibrium, then stop at the utmost position on the other side and repeat everything again... Thus oscillations will be amplified.

It's important to notice that the described oscillations, once started, will proceed without damping. Indeed, when the string moves with velocity Δv in the direction of the bow and $u > \Delta v > 0$, then the friction performs positive work. On the other hand, when going back the work of friction will be negative. The relative velocity $v_1 = u - \Delta v$ in the former case is less than that, $v_2 = u + \Delta v$, in the latter one. The friction, $F_{\text{fr}}(u - \Delta v)$ will, on the contrary, be greater in the first situation than that, $F_{\text{fr}}(u + \Delta v)$, in the

---

[a]Remember that at the linear piece of the curve, Figure 8.3, the force of friction is equal in magnitude to the resultant of the tension forces, Figure 8.2.

second one. Thus, the positive mechanical work done by friction when the string and bow are moving together surpasses the negative work performed when the string moves back. This results in the positive net work during the vibration cycle. Consequently, the amplitude of vibrations increases with each successive oscillation. And this keeps going up until a certain limit. If $v > v_0$ so that the relative velocity of bow and string v finally leaves the descending part of the plot $F_{fr}(v)$ (Figure 8.1) then the negative work of friction can overcome the positive one, forcing the amplitude of oscillations to wane.

As a result, a stable vibration with some equilibrium amplitude will finally be attained, for which the total work done by friction will be exactly equal to zero. (To be precise, the positive work during the cycle compensates the energy loss due to the air resistance, nonelastic character of deformation, etc.). These steady oscillations of the violin string will proceed without damping.

It's quite common that sound vibrations are excited when one body moves along the surface of another: dry friction in a door hinge causes it to screech; and so do our shoes, floor tiles, and so on. You can produce screeching by just pressing and pulling your finger along, say, a smooth and firm enough surface[b]. And phenomena which occur in these examples may be very similar to the excitation of vibrations of the violin string. At first there is no sliding, then an elastic deformation develops up to the point when the "take-off" happens and "their majesty" oscillations commence. And once having started, they don't subside abruptly but continue without noticeable damping. Because of the same dropping character, the friction forces do positive mechanical work that procures the energy required for the oscillations.

If the dependence of friction on relative velocity of the surfaces changes its character, the screech goes away. Everyone knows, for instance, that you can simply lubricate the surfaces to get rid of an irritating screak. And the physical reason behind it is trivial: for slow motion the liquid friction is proportional to velocity and, hence, the conditions required to bring about and then to sustain oscillations disappear when you substitute dry friction with the liquid one. Inversely, when vibrations are desirable, the participating surfaces are often specially treated in order to reach a sharper decrease of friction force with increasing velocity. For instance, for

---

[b]Chapter 9, "The chiming and silent goblets" gives a less trifling example of the kind.

this exact reason rosin is applied on violin bows.

No surprise that understanding the laws of friction often helps in solving various practical and industrial problems. For example, while machining a metal piece, undesired vibrations of the cutter can develop. These vibrations are caused by the force of dry friction between the tool and metal shaving slithering along its surface, Figure 8.4. Here again the dependence of friction on speed may have for high quality steels the familiar "dropping" character. Which, as we know by now, is the principal condition for exiting oscillations. A common way to preclude such vibration (which can turn out to be quite detrimental for both the cutter and the piece in work) will be to use along with the naive lubrication a special sharpening of the cutter, basically to hone it to the correct angle, so that no slithering will occur and there will be no reason for oscillations.

Fig. 8.4: Vibrations of cutter of machine tool may be eliminated by correct choice of the sharpening angle.

Can you describe (or even write a formula for) the motion of a string subject to constant sliding friction?

Chapter 9

The chiming and silent goblets

The carriage resembled an open shell made of glittering crystal; its two large wheels seemed to be built of the same material. When they were turning, they produced marvelous sounds: Full, yet still growing and approaching, these chords reminded the tones of glass harmonica, yet of amazing strength and power.

E. T. A. Hoffmann, *"Genannt Zinnober."*

It's not a novel idea that one can make simple wine-glasses sing. However, it turns out that there is a very peculiar way to do this. How peculiar? Well, judge for yourself.

If you dip a finger in water and start circling it carefully along the edge of a glass keeping the rim wet, it makes, at first, a rather screeching sound. But then, after the water has thoroughly and uniformly covered the glass edge, the tone should turn into something more melodic. By varying the pressure of the finger you can easily change the pitch of the produced sound. The height of the pitch will also depend on the size of the glass and thickness of its walls[a].

Notice, by the way, that not every single glass is capable of making these pleasing chiming tunes, so the search for a suitable one may turn out to be quite a meticulous affair and take a while. The best "singers" turn out to be very thin-walled goblets, those having the shape of the paraboloid of revolution, with a long slim stem. Another critical parameter

---

[a]The mechanism of exciting sound is the same as in the bow-instruments, see Chapter 8, "Why the violin sings."

that determines the resonant tone of the glass is the level of liquid in it: generally, the fuller the glass the lower its pitch. When the water level passes the midline of the glass, waves will develop on the surface of liquid, because of the wall shaking. The maximal disturbance marks the position of the finger inducing the sound at the moment.

A famous American scientist (as well as one of the greatest statesmen in the history of his country – a rare yet proven possible, at least back in those days, combination – *D. Z.*) Benjamin Franklin, who is mostly known for his experiments with atmospheric electricity[b], had employed the phenomenon discussed in the previous paragraph to create a peculiar musical instrument, very similar to that described in Hoffmann's "Genannt Zinnober". That was a series of perfectly polished glass cups, each with a drilled orifice in the middle, arranged equidistantly on the same shaft. There was also a pedal under the box where the cups on the axle were situated, like in an old-fashioned sewing machine, to make the shaft rotate. And just by simple touching with wetted fingers, one could change the tone of the system from a sound forte down to meager whistling. Now it is hard to believe, but the people who had heard this "goblet organ" playing assured that the harmony of its sounds was amazingly appeasing on both the listeners and the performer. In 1763 Franklin had given his own instrument as a present to an Englishwoman, Ms. Davis. She demonstrated it for several years traveling around Europe, and then the famous instrument disappeared without a trace. Probably memories of that true story had reached the German writer E. T. A. Hoffmann. He was a talented musician himself, and used them in his "Genannt Zinnober".

And since we are talking glasses, it seems also worth mentioning another interesting fact. It may sound iconoclastic but it's not accepted as proper etiquette to clink champagne glasses. Really. And the deal here is that for some, of course, purely physical reason, goblets filled with champagne or any fizzy carbonated drink make, when clinked, an inexpressive muffled sound. So what's the matter here? Why don't goblets with champagne ring?

Physically speaking the melodic ear-pleasing tingle that we hear after clinking glasses comes from high-frequency sound waves. Our glasses act like resonators in the high-frequency sound (that is $10-20\,KHz$) and even ultrasound (higher than $20\,KHz$) range. When we clink either empty glasses

---

[b]B. Franklin, (1709–1790), American public official, writer, and scientist.

or glasses filled with noncarbonated beverage, these oscillations keep ringing for a rather long period of time once induced. On the other hand, this automatically suggests a likely cause for muffling the sound produced by goblets with champagne. Those are the tiny pinching bubbles of carbon dioxide stampeding from the opened bottle in their delectable effervescent rush. They may lead to strong scattering of short sound waves in the goblet. Similar processes take place in the atmosphere, where fluctuations of molecular density scatter the light in the short wave part of the spectrum (see Chapter 4, "In the blue").

Even for the highest frequencies perceived by human ear ($\nu \sim 20\,KHz$) the wavelength of sound in water, $\lambda = c/\nu \sim 10\,cm$ ($c = 1450\,m/s$ is the velocity of sound in water) is considerably larger than the size of CO_2 bubbles in champagne (say, less than $1\,mm$), and, consequently, the latter seem to be quite legitimate candidates to cause a Rayleigh-type scattering of sound. Yet, let's look at the problem a bit closer. What does for instance our estimate for λ_{min} really mean? Just for simplicity, let's forget for now about the complicated shape of a real goblet and think of a rectangular box with a plane one-dimensional expansion–compression sound wave in it. For the excess of air pressure in the box we can write:

$$P_e\,(x,\,t) = P_0\,\cos\left(\frac{2\pi\,x}{\lambda} - \omega\,t\right),\qquad(9.1)$$

where P_0 is the amplitude of pressure oscillations, ω is the sound frequency, λ is the corresponding wavelength and x is the coordinate along the axis of propagation.

Since even the minimal value of λ is well over the glass dimensions, $x \ll \lambda_{min}$, the function $P_e(x, t_0)$ (called also the *pressure field* of the sound wave) at any given instant t_0 varies only slightly over the volume of the glass. Thus, the first term in (9.1) is vanishingly small and the space-time distribution of the excess pressure is determined mainly by the second term in the argument of the cosine. This actually shows that because of the negligible value of $x \ll \lambda$ the excess pressure field in the glass is almost uniform but rapidly changing:

$$P_e\,(t) = P_0\,\cos\omega\,t,\qquad(9.2)$$

Note the difference between the pressure field (9.2) and a conventional standing wave in a rigid box of length l. The resonance condition for the latter would read: $l = \frac{n\lambda}{2}$, where $n = 1, 2, \ldots$ This long sound wave simply wouldn't fit in the glass. However walls of a real glass are elastic and take part in the oscillations of the content. Vibrations of the walls transfer sound to the ambient air making it audible[c].

The total pressure of liquid in the glass is the sum of $P_e(t)$ and the atmospheric pressure:

$$P_e(t) = P_\text{atm} + P_0 \cos \omega t.$$

We are now just one step away from understanding the true reason behind the observed fast damping of sound in champagne goblets. The answer is hiding in the fact that gas-saturated liquid is a so-called *nonlinear acoustic medium*. This piece of "scientific" vernacular means in reality the following. It's known that solubility of gas in liquid depends on pressure. The higher the pressure the greater volume of gas can be dissolved in a unit volume of liquid. But, as we have already established, in the presence of sound oscillations the pressure field in the glass varies. At the moments when the pressure of liquid drops below the atmospheric one, the outgassing increases. The release of gas bubbles distorts the simple harmonic time dependence of pressure. It is in this particular sense that one calls the gas-saturated liquid a nonlinear acoustic medium[d]. The outgassing inevitably takes energy from sound oscillations making them wane much faster. After the glasses have been clinked, all kinds of sound frequencies are excited in them; then, however, due to the mechanism sketched above, high pitch-modes will die away much quicker than low-pitch ones. In the absence of the high frequencies a pure high-pitched ringing melody of crystal will turn into a miserable muffled thump.

Yet, it turns out that gas bubbles not only damp sound waves in liquid but in certain circumstances can also generate them. Indeed, recently it has been found that sound oscillations can be excited by wee air bubbles in water hit by high-power laser pulse. The effect is caused by the impact of the laser beam upon the bubble surface. There the light may experience the total internal reflection. After such a "hit" the bubble quivers for some time (until the vibration is damped) exciting sound waves in the surrounding medium. We may try to evaluate the frequency of these oscillations.

There are lots of important phenomena in nature, which, however differ-

---

[c]The original crystal voice of empty wine-glasses proves that what we usually enjoy is not the solo of liquid content but rather its duet with the splendid vessel. —A. A.

[d]Remember nonlinear sound distortions as the nightmare of Hi-Fi fans.

ent they may seem, can be described by the same equation, the equation of the harmonic oscillator. These are different kinds of oscillations – a weight bouncing on a spring, atoms vibrating in molecules and crystals, electrical charge flowing back and forth from plate to plate of the capacitor in LC-contour, and many others. The imperative physical feature that unites all the previous examples is the presence of a "restoring force". The latter occurs if the system has been driven away from equilibrium by some external perturbation. It always tends to bring the system back to the equilibrium and linearly depends on displacement. And gas bubbles vibrating in liquid are just one more example of such oscillatory systems. Hence, we can try to use the well-known relation obtained for a mass on a spring to estimate the typical frequency of oscillations of the bubbles. Of course to do so we have to figure out what plays the role of the "coefficient of elastic force" in this case[e].

The first candidate could be the surface tension σ of the liquid: $k_1 \sim \sigma$; at least it has the desired dimension (N/m). It seems reasonable to substitute into the formula for the oscillator's frequency the mass of the liquid involved in the oscillations instead of mass of the weight. Clearly this mass by the order of magnitude should be about the mass of water expelled by the bubble. This gives the volume of the bubble times the density of the liquid: $m \sim \rho\, r_0^3$. So, one can write the following expression for the natural frequency of the bubble oscillations:

$$\nu_1 \sim \sqrt{\frac{k_1}{m}} \sim \frac{\sigma^{\frac{1}{2}}}{\rho^{\frac{1}{2}}\, r_0^{\frac{3}{2}}}.$$

However it turns out not to be the only possible solution. We haven't yet utilized another important parameter, that is the air pressure inside our bubble, P_0. When multiplied by the radius of the bubble, it also gives the same (N/m) dimension of "elasticity coefficient", $k_2 \sim P_0\, r_0$. And after having plugged this new coefficient into the same relation for the natural frequency of the oscillator, we obtain an entirely different value for the

---

[e]Coefficient of elastic force k of a spring characterizes the proportionality between the value of restoring force F and displacement x from the equilibrium:

$$F = -k\,x.$$

frequency of the bubble's vibration:

$$\nu_2 \sim \sqrt{\frac{k_2}{m}} \sim \frac{P_0^{\frac{1}{2}}}{\rho^{\frac{1}{2}} r_0}.$$

Which of these two values is the true one? It may sound surprising, yet, actually both of them are correct. They just correspond to two different types of oscillation of the air bubble. The first one represents oscillations occurring when the bubble was originally squashed, say by the laser impulse. In such a motion, the shape and, therefore, the surface area of the bubble are constantly changing, while the volume remains the same. In this process the "restoring force" is determined by the surface tension[f]. The second type, on the contrary, takes place when the bubble has been squeezed uniformly from all directions and then released. In this case it starts throbbing due to the pressure forces. Our second frequency ν_2 corresponds to radial oscillations of this kind.

Obviously the effect of laser beam impact on the bubbles is asymmetrical. Therefore, the sound waves produced by the bubbles after this type of excitation are likely to belong to the first type. Further, if, for instance, the size of bubbles is known, one could determine the type of vibration from the frequency of the sound generated by the bubbles. In the experiments that we are talking about, this frequency was found to be $3 \cdot 10^4 \, Hz$. Unfortunately, the dimensions of tiny air bubbles in water are hard to measure with a sufficient degree of accuracy. It's clear though, that they should be of the order of some fractions of a millimeter. After plugging $\nu_0 = 3 \cdot 10^4 \, Hz$, $\sigma = 0.07 \, N/m$, $P_0 = 10^5 \, Pa$, $\rho = 10^3 \, kg/m^3$ into the corresponding formulae one may find the characteristic dimensions of bubbles that produce

---

[f]Notice there is a whole bunch of different kinds of oscillation, in which the bubble's volume doesn't change. Those vary from trivial alternated squashing and squeezing of the bubble in various directions to much more outlandish transformations when the bubble turns into something like, say, a doughnut. The frequencies of these oscillations may variegate quantitatively yet by the order of magnitude all of them are equal to

$$\nu_1 \sim \frac{\sigma^{\frac{1}{2}}}{\rho^{\frac{1}{2}} r_0^{\frac{3}{2}}}.$$

sound for both types of oscillations:

$$r_1 \sim \frac{\sigma^{\frac{1}{3}}}{\rho^{\frac{1}{3}} \nu_0^{\frac{2}{3}}} = 0.05 \, mm;$$

$$r_2 \sim \frac{P_0^{\frac{1}{2}}}{\rho^{\frac{1}{2}} \nu_0} = 0.3 \, mm.$$

It turns out that the size of bubbles does not differ much. Obviously, the difference is not enough to determine what type of oscillations were generated in the experiment. However, the estimates of the radii of bubbles are in perfect agreement with what we would expect from our day-to-day observations. This is evidence in favor of our way of reasoning, based mainly on dimensional arguments.

Do you have any idea why in the champagne glass the high-pitch overtones die out far quicker than the principal mode?

Chapter 10

The bubble and the droplet

The numerous guises of surface tension in the natural and technological world around us are pervasive and amazingly varied. It gathers water into droplets and allows one to blow up a soap-bubble shimmering with rainbow colors or to write with an ordinary pen. Surface tension also plays a significant role in the physiology of the human body. It's been utilized in space technology too. Why, after all, does the surface of a liquid behave in the way it does, like a stretched elastic membrane?

The molecules in the narrow layer, really close to the liquid surface, could be considered as "dwelling" in very special circumstances. They happen to have neighbors, identical to them as molecules, only on one side, whereas the "inner" denizens are completely surrounded by their twin relatives that look and act identically.

Because of the attractive interaction between these closely lying molecules, the potential energy of each of them is negative. The absolute value of the latter, on the other hand, could at first approximation be premised as proportional to the number of the nearest neighbors. It's clear then, that surface molecules, each having fewer neighbors right next to them, must have a higher potential energy than the ones in the volume of the liquid. Another factor raising the potential energy of the molecules in the surface layer is that the concentration of molecules in the liquid decreases near the surface.

Of course, molecules of liquid are moving with incessant thermal motion – some of them dive inside, leaving the surface, and others go up to take their place. However, one can always speak of the average surplus potential energy of the surface layer.

The reasoning above shows that in order to extract a molecule from inside the liquid up to the surface, external forces must perform positive work. Quantitatively this work is expressed by the surface tension σ, which is equal to the additional potential energy of molecules occupying a unit surface area (compared to the potential energy these molecules would have if they remained inside).

We know that the most stable state, among all possible states of a system, is that with the lowest potential energy. In particular liquid will always try to assume the shape corresponding to the minimal surface energy for the given conditions. This is the origin of surface tension, which always tries to shrink the surface of liquid.

10.1 Soap-bubbles

Paraphrasing the great English physicist Lord Kelvin[a], you can simply blow up a soap-bubble, stare at it, study it all your life long, and still be able to extract more lessons of physics from it. For instance, the soap film is an excellent object for exploring the various effects of surface tension.

Gravitational forces do not play any noticeable part in the considered case, for the soap film is very thin and, therefore, its mass is negligible. So the protagonist here will be the surface tension force which, as we've just shown, will try to make the surface area of the film as small as possible, within the given circumstances, of course.

But why necessarily soap films? Why can't we, for example, study, say, films of distilled water? Especially considering the fact that its surface tension is several times that of the aqueous soap solution (just a fancier name for the soapy water).

It turns out that the answer does not depend so much on the value of the surface tension coefficient, but lies rather in the structure of the soap film itself. Indeed, any soap is abundant with so-called surface-active agents *(surfactants)*, or long organic molecules with two ends that have a completely opposite affinity to the water. That is, while one end (called "the head") clings to water avidly, the other one ("the tail") stays completely indifferent to the water. This leads to the rather complex structure of soap film in which the soapy solution is armored by a fence made of those densely packed

[a]See page 53.

Fig. 10.1: Stability of soap film is guaranteed by presence of the surface-active organic molecules.

highly oriented layers of surface-active agents[b], Figure 10.1.

But for a moment, let's go back to our soap-bubbles. Most of us not only just marveled at these gorgeous creations of nature at one occasion or another, but were making them ourselves. They are so perfectly spherical in their shape and can hover in the air for so long, before rupturing finally against an obstacle. The pressure inside them appears to be higher than the atmospheric one. This additional pressure is due to the fact that the soap film of the bubble attempts to minimize the surface area and gives the air inside an extra squeeze. Moreover, the smaller the bubble radius R is, the higher the additional pressure inside. Now we shall try to find the magnitude of this addition, ΔP_{sph}.

Let's conduct a so-called mental experiment. Suppose that the surface tension of the film of the bubble drops a tiny bit, and its radius increases, consequently, by a certain value, $\delta R \ll R$, Figure 10.2. This, in turn, causes the following increase of the external surface area:

$$\delta S = 4\pi \left(R + \delta R\right)^2 - 4\pi R^2 \approx 8\pi R \, \delta R;$$

($S = 4\pi R^2$ stands for the surface area of the sphere). And, therefore, for the incremental surface energy, one can write:

$$\delta E = \sigma \left(2 \, \delta S\right) = 16\pi \, \sigma \, R \, \delta R, \tag{10.1}$$

(since δE is proportional to the tiny δR, the surface tension coefficient, σ, can be assumed to be constant).

By the way, notice the extra "2" factor showing in (10.1), although absent from the original definition of surface energy. That is because now

---

[b]Surfactants are mainly used in order to reduce the surface tension and improve wetting properties of detergents. In the meantime they help to stabilize the film and prolong the lifetime of soap-bubbles. —A. A.

Fig. 10.2:
Infinitesimal
inflation of
soap-bubble.

we have taken into account both surfaces of the bubble, its internal surface
as well as the external one; when the bubble radius grows by δR each of its
surfaces stretches by additional $8\pi R\,\delta R$.

This fictitious increment of surface energy is thanks to the mechanical
work of the compressed air trapped inside the bubble. The pressure in the
bubble remains almost the same when its volume grows by a small amount
δV so one can equate this work to the enhancement of the surface:

$$\delta A_{air} = \Delta P_{sph}\,\delta V = \delta E.$$

The volume change here, on the other hand, is equal to the volume of the
thin-walled spherical shell, Figure 10.2:

$$\delta V = \frac{4\pi}{3}(R + \delta R)^3 - \frac{4\pi}{3}\,R^3 \approx 4\pi R^2\,\delta R,$$

entailing,

$$\delta E = 4\pi R^2\,\Delta P_{sph}\,\delta R.$$

Now compare this expression to the earlier established formula (10.1).
This gives for the additional pressure inside the spherical soap-bubble that
balances the surface tension forces:

$$\delta P_{sph} = \frac{4\sigma}{R} = \frac{4\sigma'}{R} = 2\sigma'\,\rho; \qquad (10.2)$$

(we denoted by $\sigma' = 2\sigma$ the doubled coefficient of surface tension of the
liquid).

Obviously, in the case of a single curved surface (for instance, that of a spherical droplet), this additional pressure would be $\delta P_{sph} = 2\sigma / R$. This relation is called the *Laplace formula*[c]. The reciprocal of radius is conventionally called the *curvature* of the sphere: $\rho = 1/R$.

Thus we have arrived at the important conclusion that the incremental pressure is proportional to the sphere's curvature. Yet a sphere is not the only shape a soap-bubble can take. Indeed, having placed the bubble between two hoops[d], for example, one can easily stretch it in a cylinder crowned by round spherical "caps", on its top and bottom as in Figure 10.3.

Fig. 10.3: With the help of wire frames one can make a cylindrical soap-bubble.

What will the value of the additional pressure be for such an "unorthodox" bubble? It's clear that the curvature[e] of the cylindrical surface varies in different directions: it is zero along the generating line (for cylinders that is straight), however for the cross section perpendicular to the axis, the curvature equals $1 / R$, where R is the radius of the cylinder. Well, then what value of ρ should we substitute into the previously derived formula? It turns out that the difference in pressures on the two opposite sides of an arbitrary surface is defined by the average curvature. Let's try to figure

[c]Pierre-Simon Laplace, (1749–1827). A great French mathematician, one of that splendid constellation of French mathematicians, contemporaries of the Great French Revolution – J. L. Lagrange, L. Carnot, A. M. Legendre, G. Monge, etc. Laplace became best known for his contribution to the theory of probability and celestial mechanics, as well as for the famous quote from Napoleon that he "carried the spirit of the infinitely small into the management of affairs", when the scholar had miserably failed in his short-lived assignment as "Minister of the Interior".

[d]Before touching a bubble you must dip the hoops into the soap solution. —A. A.

[e]What is the curvature of a plane (two-dimensional) curve? For a circumference it is defined in the same way as for the sphere: $\rho = 1 / R$, where R is the radius. Each tiny piece of any other curve can, just the same, be considered as an arc of a certain radius. The reciprocal of this radius is called the curvature of the plane curve at the given point.

out what it will be for our right cylinder.

First, erect a normal[f] to the cylinder surface at a point A, and then construct a set of planes passing through the normal. The resulting cross sections of the cylinder by these planes, (called the *normal sections*), can be circular, elliptical or even degenerate into two parallel straight lines, (Figure 10.4). Surely, their curvature at the given point is different: it is maximal for the circle and minimal (nil actually) for the longitudinal section. The average curvature is defined then as the half sum of minimal and maximal values of the curvature of the normal sections at this point:

$$\bar{\rho} = \frac{\rho_{max} + \rho_{min}}{2}.$$

This definition applies not only to a cylinder. In principle, the average curvature at a given point can always be calculated in this manner.

For the lateral surface of a cylinder, its maximal curvature at any point is $\rho_{max} = 1/R$, where R is the radius of the cylinder and the minimal value $\rho_{min} = 0$. Thus, the average curvature of the cylinder is $\bar{\rho} = 1/2R$, and the additional pressure inside the cylindrical bubble is:

$$\Delta P_{cyl} = \frac{\sigma'}{R}.$$

So it turns out that the additional pressure in the cylindrical bubble is equal to that in the spherical bubble of the doubled radius. That makes the radii of the spherical caps of such a cylindrical bubble twice the radius of the cylinder itself. Hence the caps are just spherical segments rather than the full hemispheres.

And what would happen if one eliminates the additional pressure in the bubble completely by, say, pricking its caps? The first solution popping to mind would be that, since there is no additional pressure, the surface shouldn't have any curvature at all. Yet surprisingly, the walls of the cylinder are actually bending inside taking the shape of a *catenoid* (from the Latin *catena* for "chain"). This shape can be generated by rotating the so-called *catenary line,* around its X-axis[g]. So what's the matter here?

[f]That is, the line perpendicular to the tangential plane at the point A. —D. Z.

[g]Catenary is the curve formed by a perfectly flexible uniform chain suspended by the endpoints. The form of the curve (up to similarity transformations $a \to \alpha a$) is given by the equation:
$$y = \frac{a}{2}(e^{\frac{x}{a}} + e^{-\frac{x}{a}}).$$

Fig. 10.4: Normal
sections of cylinder
differ in curvature.

Let's examine this surface closer, in Figure 10.5. It's easy to notice that
its narrowest, or waist, portion, also called the *saddle,* is concave as well as
convex. Its section across the rotation axis is obviously a circumference; on
the other hand, dissecting along the axis gives, by definition, the catenary.
The inward curvature should raise pressure inside the bubble, but the opposite curvature would lower it. (Pressure under a concave surface is higher
than the pressure above it.) In the case of catenoid, the two curvatures are
equal in magnitude but have the opposite directions, therefore, negating
each other. The average curvature of this surface is zero. Hence, there is
no additional pressure inside such a bubble.

Fig. 10.5: Left to itself the soap film takes the form of catenoid. This surface has zero average curvature.

Catenoid is not unique though, and there are a bunch of other surfaces, seemingly "badly" curved in all possible directions, yet having their average curvature equal to zip, and consequently, not exerting any extra pressure. To generate these surfaces, it's enough to immerse a wire frame into soapy water. While lifting the frame back from the solution, one can immediately see various surfaces of zero curvature, formed depending on the frame shape. However, catenoid is the only surface of revolution[h] (besides the plane, of course) with zero curvature. Surfaces of zero curvature bounded by a given closed curve may be found with the help of methods of a special branch of mathematics, called *differential geometry*. An exact mathematical theorem claims that surfaces of zero curvature have the minimal area among all the surfaces with the same boundary – a statement which seems pretty natural and obvious for us now.

A plethora of combined soap-bubbles makes froth. In spite of seeming disorder there is an indisputable rule held in the embroidery of soap films in a foam: the films intersect one another only at equal angles, (Figure 10.6). Indeed, look, for example, at the two joint bubbles partitioned by the common wall, (Figure 10.7). The additional (with respect to atmospheric) pressures inside the bubbles will be different. According to the Laplace formula, (10.2):

$$\Delta P_1 = \frac{2\sigma'}{R_1}, \qquad \Delta P_2 = \frac{2\sigma'}{R_2}.$$

So the common wall must be bent in order to compensate the pressure difference in the bubbles. Its radius of curvature, hence, is determined by

[h]That is the surface which may be generated by rotation of a curve.

the expression:

$$\frac{2\sigma'}{R_3} = \frac{2\sigma'}{R_2} - \frac{2\sigma'}{R_1},$$

which gives after regrouping

$$R_3 = \frac{R_1 R_2}{R_1 - R_2}.$$

Again, Figure 10.7 depicts a cross section of these two bubbles by a plane passing through their centers. The points A and B here mark the intersections of the plane of the picture with the circumference where the bubbles are touching. At any point on this circumference there are three films coming together. As long as their surface tension is the same, the tension forces can balance each other only if the angles between the crossing surfaces are the same. Therefore, each of them is equal to $120°$.

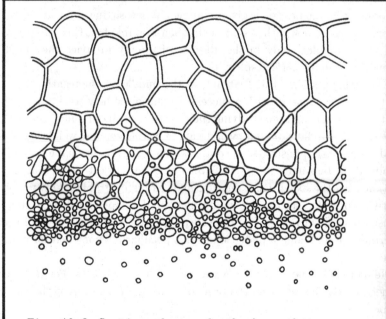

Fig. 10.6: Section of soap froth shows that joining films make equal angles.

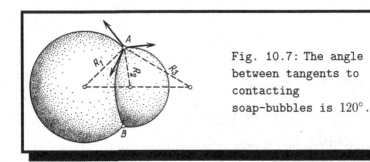

Fig. 10.7: The angle between tangents to contacting soap-bubbles is 120°.

10.2   On different kinds of droplets

The shapes of droplets. Here things become a bit more complicated. Now the surface tension which, as always, tries to minimize the area of the surface is counteracted by other forces. For example, a liquid droplet is almost never spherical, although it is a sphere that has the smallest surface area among all shapes for a given volume. When sitting on a flat steady surface, droplets look rather squashed; when in a free fall, their shapes are even more complex; only in the absence of gravity in space they finally assume the form of perfect spheres.

The Belgian scientist J. Plateau[i], in the middle of nineteenth century, was the first to come up with a successful solution of how to eliminate the effects of gravitation when studying the surface tension of liquids. Surely, in those days researchers did not even dream of having genuine weightlessness, and J. Plateau simply suggested compensating the gravitational forces with Archimedean buoyancy force. He submerged his subject liquid (oil) in a solution with the exact same density, and, as his biographer tells us, was utterly surprised to see that the oil droplet had developed a spherical shape. He used his golden rule to "become surprised in the right time," and then experimented and contemplated upon this peculiar phenomenon for a long while.

He had his method to study a variety of entailing effects. For instance, he meticulously investigated the process of droplet formation at the end of a tube.

Normally, no matter how slow a droplet is being made, it separates from

[i] Joseph Antoine Ferdinand Plateau, (1801–1883), Belgian physicist; worked in the fields of physiological optics, molecular physics, surface tension. Plateau was the first to put forward the idea of a stroboscope.

the tip of the tube so fast that the human eye can't follow the details of this event. So Plateau had to dip the tip of the tube he was using into a liquid, with a density just slightly less than that of the droplets themselves. The gravitational force influence was, by doing so, substantially diminished and as a result, really large droplets could be formed and the process of their taking off from the end of the tube could be clearly seen.

In Figure 10.8 you see the different stages droplets undergo in their formation and separation (of course, these pictures were taken using the modern high-speed film technique). Let's try now to explain the observed sequence. During the slow growth stage, the droplet is in equilibrium at each particular instant. For a given volume, the droplet shape is determined by the condition of a minimal sum of its surface and potential energies, the latter of which is the result of gravitational forces, of course. The surface tension is trying to shape droplets spherically, whereas gravity, on the contrary, tends to situate the droplet's center of mass as low as possible. The interplay of these two yields the resulting vertically stretched form (the first shot).

Fig. 10.8: High-speed photographs of droplet take-off.

As the droplet continues to grow gravitation becomes more prominent. Now most of the mass accumulates in the lower part of the droplet, and the droplet begins to develop a characteristic neck (the second shot of Fig 10.8).

The surface tension forces are directed vertically along the tangent to the neck, and for some time they manage to balance the droplet's weight. Not for long, however, and at a certain moment just a slight increase in the droplet's size is enough for gravity to overtake the surface tension and to break this balance. The droplet's neck then narrows promptly (the third shot), and off the droplet finally goes (the fourth shot). During this last stage, an additional tiny droplet forms on the neck and follows the big "maternal" one. This secondary little droplet (called Plateau's bead) is always there; however, because of the extreme swiftness of the droplet leaving the tap, we basically never notice it.

We won't be going into the details of the formation of Plateau's bead as it is a pretty complicated physical phenomenon. We will instead try to find an explanation of the observed shape of the primary droplets in their free fall. Instant photographs of the falling droplets show clearly the almost spherical shape of the little secondary drops, while the big primary ones look rather flat, something like a bun. Let's estimate the radius at which the droplets start losing their spherical shape.

When a droplet is moving uniformly (at a constant speed), the force of gravity acting, say, on the narrow central cylinder AB of the droplet, Figure 10.9, must be balanced by forces of surface tension. And this automatically requires that the radii of curvature of the droplet at A and B should differ. Indeed, the surface tension produces the additional pressure, defined by the Laplace formula: $\Delta P_L = \sigma' / R$, and, if the curvature of the droplet surface at the point A is greater than that at the point B, then the difference of these Laplace pressures could compensate for the hydrostatic pressure of liquid:

$$\rho g h = \frac{\sigma'}{R_A} - \frac{\sigma'}{R_B}.$$

Let's check how much R_A and R_B should really differ to satisfy the previous relation. For tiny little droplets with radii of about $1\,\mu m$ ($10^{-6}\,m$), the value of $\rho g h \approx 2 \cdot 10^{-2}\,Pa$, whereas $\Delta P_L = \sigma' / R \approx 1.6 \cdot 10^5\,Pa$! So, in this case, the hydrostatic pressure is so small when compared to the Laplacian that one can safely disregard it, and the resulting droplet will be very close to an ideal sphere.

But it's a completely different story for a drop of, say, $4\,mm$ radius. Then the hydrostatic pressure is $\rho g h \approx 40\,Pa$, but the Laplacian one is $\Delta P_L = 78\,Pa$. These are the values of the same order of magnitude, and,

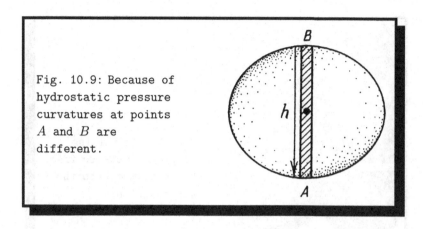

Fig. 10.9: Because of hydrostatic pressure curvatures at points A and B are different.

consequently, the deviation of such a droplet from the ideal spherical shape becomes quite noticeable. Assuming $R_B = R_A + \delta R$ and $R_A + R_B = h = 4\,mm$, one finds $\delta R \sim h \left[\sqrt{\left(\frac{\Delta P_L}{\rho g h} \right)^2 + 1} - \frac{\Delta P_L}{\rho g h} \right] \approx 1\,mm$, and the difference of the radii of curvature at A and B now turns out to be of the same order as the size of the droplet itself.

The previous calculations show us which droplets should deviate substantially from the sphere. However, the predicted asymmetry turns out to be the opposite of that observed in the experiment (indeed, real droplets in Figure 10.10 are flattened from the bottom!). What's the matter here? Well, the thing is that we assumed the air pressure to be the same over and under the droplet. And this is certainly true for slow moving drops. But when the speed of the droplet is sufficiently high, the surrounding air does not have enough time to smoothly flow around. A region of pressure appears before the droplet and an area of lower pressure forms right behind it (where real turbulent vortices are formed). The difference of the front and back pressures can actually exceed the hydrostatic pressure, and the Laplace pressure now must compensate for this difference. In such circumstances, the value of $\frac{\sigma'}{R_A} - \frac{\sigma'}{R_B}$ turns negative, meaning that R_A is now greater than R_B. That (to our final satisfaction) is exactly what has been seen in the experimental pictures.

And finally, a short quiz about the giants and whoppers. Have you ever seen them among droplets? Not many. They simply do not survive under normal circumstances. And for good reason: droplets of large radii

Fig. 10.10: Falling
droplet flattened from
the bottom due to the
air resistance.

turn out to be unstable and spatter into a bunch of little ones almost
instantly. It's the surface tension that assures longevity of a droplet on a
hydrophobic surface. Yet once the hydrostatic pressure becomes greater
than the Laplacian, the droplet spreads over the surface and breaks into
smaller ones. One can use the following relation to estimate the maximal
radius of a still stable droplet: $\rho g h \gg \frac{\sigma'}{R_A}$, where $h \sim R$. From which, one
can find:

$$R_{max} \sim \sqrt{\frac{\sigma'}{\rho g}}.$$

For water, for instance, $R_{max} \approx 0.3\,cm$ (of course, this is just the order
of magnitude estimate of the maximal size of droplets). That's why we
never see, for example, really gigantic droplets on the leaves of the trees or
other unwettable surfaces.

Chapter 11

The mysteries of the magic lamp

... "Simmetriads" appear spontaneously. Their birth resembles eruption. All of a sudden, the ocean starts coruscating as if tens of square kilometers of its surface were covered with glass. A while later, this glassy envelope pops up and bursts outwards in a shape of a monstrous bubble, in which, distorted and refracted, arise the reflected images of the whole firmament, the sun, clouds, the horizon...

Stanislaw Lem, *"Solaris."*

The series of pictures, shown on the cover, haven't been taken either on Solaris, or from a spaceship diving in the recondite abyss of Jupiter's atmosphere. Neither have they been taken from the window of a bathyscaphe having dared to approach an erupting underwater volcano. Not even close. They are just photographs of a working Lava-lamp, a gadget anyone could find without much trouble, for example, in a "hands-on" toy shop or sometimes in a big department store[a]. And yet, it turns out that this seemingly simple device conceals plenty of beautiful and subtle physical phenomena.

The design of the lantern isn't very complicated. It consists of a cylinder with transparent walls, in the base of which under its glass bottom a regular electrical bulb is mounted. The glass in the lower part is covered with a multicolor light filter, and a coiled metal wire is wrapped around the bottom

---

[a]It's going to cost you fifty bucks though. In the long desisted Soviet Union the same gadget would run you only one tenth of that. It's just one more example of how much the cost of research may vary depending on where one conducts the experiments. — D. Z.

perimeter as in Figure 11.1. One sixth of the cylinder is filled with a wax-like substance (which we will call from now on *substance* $\mathcal{A}$), and the rest of its volume is filled with a transparent liquid (say, *liquid* $\mathcal{B}$). The particular criteria for choosing these substances, as well as their properties, will be discussed a bit later, when we will be closely studying the physical processes taking place in the lamp.

It appears more convenient to conduct observations of the Lava-lamp in the dark, having it as the only source of light. So, let's turn it on and prepare to wait. As we will see, the events taking place inside the lamp could be separated into several stages. We will call the first one "the phase of rest and accumulating strength".

The substance $\mathcal{A}$ is amorphous and, therefore, does not have a strict, well-ordered internal structure[b]. As its temperature goes up, it becomes more and more malleable, softens, and gradually turns liquid. It's worth recalling at this time a principal difference between the crystalline and amorphous substances. For the former ones, this solid-to-liquid transition (melting, in ordinary words) happens at a certain temperature point and requires a particular amount of energy (*the heat of melting*), which is expended for breaking the material's crystal structure. On the contrary, solid and liquid states of an amorphous substance aren't critically different. When the temperature is rising, amorphous materials simply soften and become liquidlike.

When turned on, the bulb of the lantern, illuminating the cylinder from the underneath through the color filter with a kind red-green glow, serves also as the heater. In the bottom floor, close to the bulb, there consequently develops a "hot spot" (the area of elevated temperature). Substance $\mathcal{A}$ in this hot region becomes softer, whereas in the same time neither the upper crust of $\mathcal{A}$, nor the liquid $\mathcal{B}$ have had enough time to warm up, remaining relatively cold. As the larger part of $\mathcal{A}$ softens, the top solid crust becomes thinner. At the same time, due to the thermal expansion, the volume of the lower, now liquid, part of $\mathcal{A}$ tends to increase, raising the pressure underneath the crust. Eventually, $\mathcal{A}$ finally breaks the crust and lunges bubbling upwards. It's like a tiny volcano is born. The quiescent phase of "rest and accumulating" is over and the new period of "volcanic activity" kicks in (see top-right figure on the cover).

---

[b]We'll dwell on the difference between crystalline and amorphous substances later in Chapter 20.

Fig. 11.1:
Construction of the
magic Lava-lamp.

The substances A and B are chosen in such a way that the density of
the warmed A, rushing up from the crack, is slightly higher than that of
the still rather cold B, causing the new portions of A, successively leaving

the rift, to surface one after another[c]. On their way up, these pieces start cooling down and, when reaching surface, become solid again, assuming various and quite peculiar shapes. Yet their density is back to the initial value, which is higher than the density of B, and these smithereens begin to sink slowly. Some of them, usually the smaller ones, however, continue to hover near the surface for a long time. And the reason for this recalcitrant behavior is, of course, our old acquaintance, surface tension. Indeed, A and B are chosen so that the B-liquid does not wet the A-solid. Hence the surface tension force acting on fragments of A is directed up, trying to push them out from the liquid. It's exactly the same reason why, for example, the water-striders can freely stay (and quite audaciously run) on the surface, or an oiled metal needle does not founder.

Meanwhile, the excessive pressure in the lower part of the cylinder, under the crust, has been relieved, the edges of the crack have become molten, and new portions of melted A are continuously trickling out of the crater. However, now they don't sever in the form of bubbles, instead they stretch leisurely as an extended upward narrow stream. The outer surface of this stream, in contact with the cold B, quickly cools down and stiffens, producing a sort of trunk. And if one tries to look through this trunk, one may well be surprised, for the trunk turns out to be a hollow, narrow walled tube, filled with liquid B. The explanation would be that when the stream of melted A leaves the crater and runs upwards, at some point it simply doesn't have enough material to continue growing. Then, pressure inside the trunk decreases and, resulting from that, a crack develops somewhere in the junction between the crater edge and the trunk, and then, sure enough, the cold liquid B starts pouring into the cleft. The top of the A tube in the meantime keeps going up, and the liquid B fills the tube inside, cooling down and shaping the tube's inner walls, finally causing them to completely solidify.

As the vine of the volcanic plant makes its way to the surface, on the bottom of the lantern the melting continues and the next ball of the "hot" liquid A leaves the crater. It goes up, but now it goes up inside of the developed tube, and when it gets to the top of the tunnel, the ball, still being warm enough, extends the tube by another increment. So the plant

[c]This resembles the famous experiment, in which a droplet of aniline, that at first peace-fully rests at the bottom of a tall glass cylinder with water, immediately starts up to the surface as soon as the temperature reaches about $70°$ C and the aniline's density becomes less than that of water.

keeps on growing, adding one by one these successive blocks (see second-top figure on the cover). Soon, shoving off the crusty chips of the preceding "volcanic activity", there starts protruding another stem near the first one, and maybe another, after. These esoteric underwater plants swirl and intertwine, like exotic shoots of jungle verdure, among the falling rocks, continuing at leisure to descend from the surface. The picture halts for a time. We could name this stage "the phase of the rocky forest".

If at this point one turns the lantern off, the petrified thicket will remain there "forever" and the lamp will not be able to return to its original, clearly segregated two-phase state[d]. Yet, surprisingly, after the kaleidoscope of the described events, we have not reached the working regime of our magic lamp. So let's keep it on and continue to watch.

In time, the liquid B is still warming up, and the boulders resting on the floor are starting to melt again, the tangled vines of the magnificent plants are wilting down. An interesting fact: there are no really squeezed shapes among the droplets formed from the liquefied rocks. They all are turning out quite spherical. Under normal conditions, the force squeezing water droplets on a hydrophobic surface is their weight. And it is balanced by the force of surface tension, tending to make the drops ideally spherical, for the sphere has the minimal surface area for a given volume. In the Lava-lamp's flask, besides gravity and the surface tension, there is an Archimedean buoyancy force also acting on the droplets, and, because of the closeness of the densities of A and B, almost equalizing the gravitational force. So the droplets happen to be in a kind of nearly weightless situation, with nothing to preclude them from donning their predestined round shape (we have already discussed that topic in Chapter 10).

For a single droplet, in absence of gravity, the ideal spherical form is the most energetically favorable one. For two, or several drops, touching each other, from the same logic, it would be more beneficial to merge into one, because the surface area of a single large ball is less than the total area of surfaces of several smaller ones, of the same aggregate mass (we will allow the reader to check this statement on their own). However, looking at a Lava-lamp, one notices that those almost spherical droplets of A tend to linger together without actually coupling. It seems especially striking if one remembers how promptly, almost momentarily, mercury or water droplets, for example, couple on an unwettable surface. What determines the time

---

[d]Unless you switch it on again. —A. A.

it takes for, say, two droplets to join?

Interestingly enough, this question attracted the attention of different researchers and engineers for a long time. Not only from the point of simple scientific curiosity, but also because of its critical importance for understanding physical processes in some very practical areas. For instance, in powder metallurgy, when the preliminary metals are powdered into grains, pressed and *baked* together to produce new alloys with desired physical properties. Back in 1944, the bright Russian physicist Yakov Frenkel[e] had proposed a simple and quite useful model of such a merging process in his pioneering work, which became fundamental in establishing the theoretical basis for this important branch of modern metallurgical technology. And now we are going to use the underlining idea of his work to estimate the time it would take for two droplets in the magic Lava-lamp to couple.

Let's consider two identical droplets in close proximity, so that they start touching each other. At the point of their contact a connecting isthmus develops, Figure 11.2, which continuously grows as the two droplets merge. We will use the energy considerations (for it is the simplest and the shortest way) to estimate the coupling time. The total energy available for the system of the two drops ΔE_s results from the difference between the surface energies of its initial and final states, that is the summed surface energy of the two separate droplets of radii r_0 and the big "unified" droplet of the radius r:

$$\Delta E_s = 8\pi \sigma r_0^2 - 4\pi \sigma r^2.$$

Since after merging the total volume of the drops does not change, one can write the following equality: $\frac{4\pi}{3}r^3 = 2 \cdot \frac{4\pi}{3}r_0^3$, and find $r = r_0 \sqrt[3]{2}$, so that

$$\Delta E_s = 4\pi \sigma \left(2 - 2^{\frac{2}{3}}\right) r_0^2. \tag{11.1}$$

According to the Frenkel idea, this additional energy is spent for work against forces of liquid friction, appearing in the process of redistributing droplet material as well as the surrounding medium, during the droplet merging. We can estimate this work by the order of magnitude. To find the

---

[e]Ya. I. Frenkel, (1894–1952), specialist in solid-state physics, physics of liquids, nuclear physics, etc. It's worth mentioning that in 1936 Ya. Frenkel had independently of N. Bohr proposed the so-called *drop nuclear model*. In this context coalescing droplets stood in direct relation to problems of nuclear synthesis. —*A. A.*

Fig. 11.2: Initial stage of the droplet coalescence.

liquid friction force, we will apply the famous Stokes' formula[f] for a spherical ball of radius R moving with velocity $\vec{v}$ in a liquid of viscosity η: $\vec{F} = -6\pi\,\eta\,R\vec{v}$. We suppose further that the viscosity of the droplet material is significantly higher than that of the surrounding liquid $\mathcal{B}$, which allows us to leave η_A in the Stokes' expression as the only viscosity coefficient[g]. Also, we may plug r_0 in the place of R. And then, noticing that the same quantity characterizes the scale of mutual displacement of the droplets when they merge: $\Delta x \sim r_0$. So finally, one could write for the work of the liquid friction forces:

$$\Delta A \sim 6\pi\,\eta_A\,r_0^2\,v.$$

From this expression, it's clear that the faster the droplets are merging the greater is the amount of energy required (because the liquid friction force increases with speed). However, the available energy resource is limited by ΔE_s, (11.1). So these two relations will provide us with the required merging time τ_F (called the *Frenkel time* of merging). Assuming $v \sim r_0\,/\,\tau_F$ to be the process speed, we find:

$$\Delta A \sim \frac{6\pi\,\eta_A\,r_0^3}{\tau_F} \sim 4\pi\,\sigma\left(2 - 2^{\frac{2}{3}}\right)r_0^2,$$

[f]George Gabriel Stokes, (1819–1903), renowned British physicist and mathematician, mostly famous for the theorem and the formula, both having commemorated his name.

[g]The Stokes' expression was derived for a different situation, when a spherical body moved in viscous liquid. However, it's pretty obvious that in the case of two merging droplets, the liquid friction force can only depend on viscosity, droplet size, and the speed of the process. Hence, from the dimension considerations, the Stokes' formula turns out to be the only combination of these three physical qualities with the dimension of force (and we do not care about the exact proportionality coefficient, for our estimate is by the order of magnitude only).

and finally:

$$\tau_F \sim \frac{r_0 \, \eta_A}{\sigma}. \tag{11.2}$$

For water droplets of, say, $r_0 \sim 1 \, cm$, $\sigma \sim 0.1 \, N/m$ and $\eta \sim 10^{-3} \, kg/(m \cdot s)$ this time turns out to be around just evanescent, $\sim 10^{-4} \, s$. Yet, for example, for the much more viscous glycerin ($\sigma_{gl} \sim 0.01 \, N/m$, $\eta_{gl} \sim 1 \, kg/(m \cdot s)$ at $20°\, C$), the corresponding time is $\sim 1 \, s$, proving the fact that for different liquids, depending on their viscosity and surface tension coefficient, τ_F can vary within a rather broad range.

It's worth emphasizing at this point that even for the same liquid, due to the strong temperature dependence of viscosity, Frenkel time can vary quite a bit. Going back to the glycerin, for instance, its viscosity drops 2.5 times when the temperature rises from $20°$ to $30°\, C$. The surface tension coefficient, on the other hand, stays pretty indifferent to the temperature variation – in the considered temperature range σ_{gl} doesn't change by more than a couple of percent. This allows us to safely assume that the temperature dependence for Frenkel time is purely determined by the viscosity temperature dependence.

Now, let's look again at the balls of $\mathcal{A}$ still lying peacefully on the floor of the lamp through the derived estimation for the Frenkel merging time. As long as the liquid $\mathcal{B}$ stays rather cold, $\mathcal{A}$'s viscosity remains low, and it will prevent the balls from rushing to merge. It is the same reason that two touching wax balls do not couple into one at room temperature. Although, if one heats them enough, the viscosity of the wax plummets and the balls merge expeditiously. One more thing that plays an important role in the process is the state of the surfaces of the potential partners – if they are rough and contaminated it is difficult for the initial bridge to develop.

The merging of the $\mathcal{A}$ droplets is absolutely critical for the lantern's working cycle to go on. This explains the presence of a special means, in order to facilitate this redistribution of $\mathcal{A}$, from numerous drops into a uniform, melted mass. It is the metal coil, wired along the lantern's bottom perimeter. This coil is warmed up by now, and when approaching and touching it, the droplets receive that needed heat, lowering their viscosity and, by doing so, warming greatly their desire to join back in the prime body of the liquefied $\mathcal{A}$. Soon, after all droplets finally disappear into the maternal mass, one, unified liquid phase of $\mathcal{A}$ remains in the bottom of the lantern cylinder. And, because it is still continuously being heated, the liquid $\mathcal{A}$ can not remain motionless. A new stage of the Lava-lamp life commences. We will call it "the phase of protuberances".

Formed in the surface layers of $\mathcal{A}$, such protuberances languidly take off for their upward trek to the surface of $\mathcal{B}$, pulled, of course, by the buoyancy force, and gradually assuming, as they go, an increasingly spherical shape (see middle figure on the cover). Having reached the upper layers of $\mathcal{B}$ (where $\mathcal{B}$, due to its low thermal conductivity, still remains cold), the protuberances cool down a little, nerveless, remaining liquid this time, and begin to drown slowly, landing on the $\mathcal{A}$'s slightly popped up surface. Because of their relatively high viscosity, it is quite difficult for them to dive into the $\mathcal{A}$ medium right away. So they bounce on its surface for a while, drifting to the periphery, where the "surgical" metal coil opens up their surface and they end their life cycle exactly where they started it.

The bulb in the cylinder base keeps heating the system, creating new protuberances. As the temperature continuously rises, the rate of their birth goes up as well. When taking off from the $\mathcal{A}$ surface, protuberances leave behind them smaller droplets[h], which kind of freeze perplexed in space, hesitating whether they really should go up into the unknown following their parent, or maybe just return to safety, into the original medium. In time, a dozen of such orphaned liquid balls are hovering in the cylinder, some of which do finally dare to continue upwards, whereas the coy ones descend back (see second last figure on the cover): a new stage of "collisions and calamities" emerges. And this turns out to be the longest and the most impressive phase of the lantern's activity.

The spheres are colliding, veering in various directions, however managing to avoid merging in the process. It seems like it would be advantageous, energetically, for the striking drops to couple (for the same reason we've mentioned just some paragraphs previously). Yet, once again, they happen to run into the time problem. The duration of collision t, is all they have, and if τ_F turns to be much longer than t, than there is not enough time for the droplets to join and having collided, they will just simply bounce apart. Let's try to give an estimate of the collision time. Most of the collisions in the lamp are glancing ones (Figure 11.3), during which the soft liquid balls slightly deform and slide along each other. The characteristic time of such an encounter must be about $t \sim r_0 / v$. The velocity of the balls flowing in $\mathcal{B}$, v, is just several centimeters per second, and the ball radii are of the

---

[h]By the way, these are the same Plateau balls that have been mentioned in Chapter 10.

order of a couple of centimeters too. It makes $t \sim 1\,s$, which is too short for them to unite, leaving then no other option except to continue roaming aloofly and unattached in the lamp cylinder. At times they loiter by the bottom, then they wander though the bulk of $\mathcal{B}$, colliding with each other, yet not merging.

Fig. 11.3: Most of collisions of $\mathcal{A}$-balls in the lamp are soft glancing touches.

This *"phase of collisions and calamities"* can go on for hours. The manual usually recommends switching the lantern off after 5–7 hours of operating. But under certain circumstances, when the temperature of the ambient air is sufficiently high (say, you happen to marvel at the magic gadget on a sultry Austin or squashingly hot Tucson summer evening), the described collision stage turns out not to be the last one. Finally, after a stationary temperature distribution along the cylinder height has been attained (and the whole liquid $\mathcal{B}$ has warmed up), the densities of $\mathcal{A}$ and $\mathcal{B}$ become practically the same, and the entire $\mathcal{A}$ congregates into a single gargantuan ball. At first, the whopper is hanging in the bottom part, bouncing at times against the cylinder walls. Then, because of these "colder wall" contacts, it cools down a little, becomes a bit denser, and, consequently, sinks down to the floor. After touching the bottom though, the ball gains an extra portion of heat, its density drops again, and it returns to its previous position, where it stays until it cools down again. Then the cycle starts all over. This phase, unmentioned in the lantern's instructions, we could name "the super-ball time" (see bottom-right figure on the cover).

Finally, after we, along with our magic Lava-lamp, have gone through the numerous stages of its work, gaining some understanding of the mechanism of the described processes, let's take a concluding look at these phenomena in a general manner. The first question coming to mind is why do these successive, often repeating, events of the birth, life, and death of the spheres occur at all? It is clear, from all our previous discourse, that the driving force behind the processes is the temperature difference between the top and bottom ends of the lantern (in thermodynamical terms, between the *heat source* and the *heat sink*). If one supposes that the flux of heat propagates in the system only because of the heat conduction by the B-liquid, the temperature of B will be simply changing gradually along the height, and nothing unusually amusing would happen. The birth of spheres, on the other hand, as well as ordinary convection, is a consequence of instabilities which sometimes develop in systems, where a thermal flow due to the variation of temperature along the boundary occurs. The study of the behavior and properties of these systems is the subject of a rather new, though quite volcanically developing branch of physics called *Synergetics*.

Chapter 12

The water mike or about one invention of Alexander Bell's

These days, everybody knows what the microphone is. Right? We see it on TV: those nifty pin-looking things clinging to the anchorman's lapels, or the old-fashioned, a ball on a handle type that reporters stick into people's faces in feverish anticipation of a story. Radio-interviewers often ask their guests to speak directly into their microphone, hinting to us the presence of the latter; and movie makers, no matter how sophisticated and outlandish their sound effects are contrived to be, end up registering them with some kind of a microphone. One can easily buy a decent mike in any home electronics store and use it with their tape or CD recorder, computer or telephone. The design of the gadget is described in most of today's high school physical textbooks. However, we can assure our revered reader that there are only a few who are aware of the existence of the so-called water microphone. Don't be surprised. Indeed it turns out that one can rather efficiently amplify different sounds with the help of a simple water stream. The device employing such a principle of sound amplification was invented by the American engineer Alexander Graham Bell, who is mostly known as one of the inventors of another gadget we can't imagine our daily life without – the telephone[a].

But first, let's pay attention to that "water stream" amplifier thing.

If there is a hole, say a little round orifice, drilled in the bottom of a reservoir of water, one may notice that the stream flowing downwards through the hole consists of two parts, differing in properties. The upper is transparent and steady, looking as if made of glass; yet as it goes

---

[a]Alexander G. Bell, (1847–1922), Scottish-born American inventor. The first public demonstration of speech transmission using his electrical apparatus took place in 1876.

further from the outtake, the stream becomes thinner and finally at the
point of the minimal cross section, it turns into the second part, which is
rather opaque and jittery. At first glance, it still looks continuous, without
interruptions, like the upper region. However, it is possible at times to
swiftly pass one's finger across this part of the stream without getting wet.
The French physicist Felix Savart[b] after having meticulously investigated
properties of liquid streams, arrived at the conclusion that at the narrowest
point of the stream it breaks its continuity and splits into series of separate
droplets. Today, over a century after the discovery, one can easily prove
this by taking photographs of the trickle with a flash, or by looking at the
stream in stroboscopic lighting, see Figure 12.1; in Savart's day, however,
researchers had to study the trickles in the dark, observing them with light
from electrical sparks.

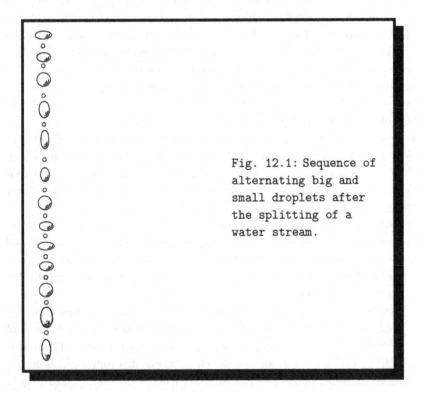

Fig. 12.1: Sequence of
alternating big and
small droplets after
the splitting of a
water stream.

[b]F. Savart, (1791–1841), French physicist; worked in the fields of acoustics, electromag-
netism and optics.

Look at the momentary image of the lower portion of the stream, Figure 12.1. It's composed of successive, alternatively bigger and smaller, droplets. As the picture clearly shows, the bigger ones are actually oscillating, gradually changing their shape from a flattened, horizontally stretched ellipsoid (droplets 1 and 2 in the picture), to round balls (3), then to the ellipsoid again (4, 5, 6), squeezed and stretched vertically, and then back to sphere (7), etc. Each droplet, pulsing rapidly in its free fall[c], produces different images in one's eye at different instants. This causes us to perceive the lower part of the stream as kind of misty, widening in the regions where the droplet-ellipsoids are stretched latitudinally, and narrowing where they are elongated vertically.

Another interesting finding of Savart's was that sounds have a strong effect on the upper transparent part of the water jet: if a sound of a certain pitch (meaning frequency) is excited nearby, the transparent region of the stream turns instantly opaque. Savart gave the following explanation. The droplets, which the stream finally breaks into, actually start developing from the very beginning of the fall, right by the outlet. At first they are outlined only by circular notches that become more prominent as the liquid falls, until the point in the stream where they split completely. These notches are so close together that they make a slight sound. Thus a music tone, in unison with this "natural" pitch, will make the continuous stream of liquid break into separate droplets earlier turning the transparent flow bleary.

The English physicist John Tyndall[d], repeated later F. Savart's experiments in his laboratory. He managed to produce a water stream, transparent and uninterrupted, of about 90 feet long. And then, by using the sound of appropriate tone and volume from one of the pipes of an organ, he was able to break this stream into countless separate drops, transforming it by doing so into a misty unsteady trickle. In one of his articles, he re-

---

[c]The frequency of pulsations can be estimated similarly to that of air bubbles in liquid, Chapter 9, "The chiming and silent goblets":

$$\nu \sim \sigma^{1/2} \, \rho^{-1/2} \, r^{-3/2}.$$

Putting $\sigma = 0.07 \, N/m$, $\rho = 10^3 \, kg/m^3$, $r = 3 \cdot 10^{-3} \, m$, one finds that $\nu \approx 50 \, Hz$. Worth noting here is that the "shooting rate" of a typical movie camera is 24 pictures per second. This is enough for the human eye to take the film for continuous motion.

[d]J. Tyndall, (1820–1893), English physicist, specialized in optics, acoustics and magnetism.

lated his observation of the water stream falling into a basin. He observed that the water stream falling into a basin sounds like that when the falling stream crossed the surface above the point of transition from the transparent to opaque part at moderate pressure, then the stream entered the liquid silently; but when it crossed the surface below the rupture point, a murmuring started and one saw numerous bubbles formed. In the former case, not only was there no serious sputtering, but liquid rather piled up around the basis of the stream in the basin where the direction of the liquid was actually reversed.

These features of water streams were used by A. Bell in his design of the water mike, depicted in Figure 12.2. It consisted of a metal tube and a branch pipe with a funnel soldered on its side; the bottom of the tube was mounted on a massive support, whereas its top was covered with a membrane-like rubber piece, fastened to the tube by a lace. As we know already from Tyndall's experiments, the lower part of the water stream, split into separate droplets, makes a murmuring noise as it reaches the water in the basin. On the other hand, when the upper, still continuous, portion of the stream enters the liquid in the reservoir, the flow remains soundless. One can perform a similar demonstration with a piece of cardboard, placed across a stream of water. As one gradually pulls the cardboard sheet up, the drumming generated by hitting droplets becomes lighter and, after the "transition point" is passed, the noise ceases.

The membrane in Bell's microphone plays exactly the same role as the cardboard piece in the preceding example. Yet, because of the resonator (tube) and the side pipe with the trumpet, the slightest tap of a droplet is amplified and echoed much louder. Thus, the tiny droplets that hit the rubber membrane will make a rapping equal to that of a hammer against an anvil.

One could easily use this apparatus to illustrate the sensitivity of a water stream to different musical tones, the fact described by Savart and Tyndall. That is, if we touch a faucet from which a slim stream of water runs by a vibrating tuning fork, Figure 12.2, the flow will momentarily break into drops which "commence" their rather earsplitting chorus. This amplification of the original sound which was pretty weak at the expense of the energy of the falling stream does indeed constitute the physical principle of the water microphone. If one substitutes the tuning fork with, say, a wrist watch, it will make its ticking audible to the entire audience in the room. One of the renowned popularizers of science at the end of nineteenth century

Fig. 12.2: The water microphone by Alexander Bell amplified sound at the expense of energy of falling droplets.

claimed that he had tried to transmit the sound his voice by connecting a funnel to a glass tube from which water was running. Although the water stream in his utensil had presumably started "talking", the voice was so roaring, horrifying and indiscernible, that according to the legend, the spectators just scooted away[e]. While reading these lines, the authors feel that it is quite a blessing that Bell's main invention, the telephone with the electrical mike in its receiver, happened to be free of such a disadvantage.

Returning to Chapter 10, "The bubble and the droplet", try to understand why droplets in the lower part of the stream are pulsing periodically. Are the small ones throbbing too?

---

[e]See footnote on page 70.

Chapter 13

How the waves transmit information

> It is monstrous, monstrous:
> Methought the billows spoke and told me of it;
> The winds did sing it to me, and the thunder,
> That deep and dreadful organ-pipe, pronounced
> The name of Prosper: it did bass my trespass.

> W. Shakespeare, *"The Tempest."*

We have grown so accustomed to television, radio, cellular phones, and Internet that we are no longer even a tiny bit surprised by the fact that we can so easily receive information from pretty much any corner of the world. Yet it was not always like that (actually not even close) and not a very long time ago...

As Russian authors, we may just as well use an example from Russian history. In order to send a message to St. Petersburg about the coronation ceremony of Queen Elizabeth which took place in Moscow in 1741, a human chain of soldiers with signal flags in their hands was mustered all along the road from St. Petersburg to Moscow. When the crown was put on the new empress' head, the first of the soldiers waved his flag, so did the second when he had seen his neighbor signaling, then the third, the fourth, and so on. So the news about the crowning event had made its way to the Northern Russian Capital, where a fired cannon notified the crowds of anticipating subjects.

Now, let's ask ourselves a legitimate question: what was actually "moving" along this peculiar chain? Although each soldier remained in his original place, at a certain moment he changed his *state,* raising the flag. And this change of state was exactly the thing which moved along the chain. In situations of this sort physicists say that a *wave was running (propagating)* along it.

There are a great variety of waves, depending on what physical properties vary when they propagate. For sound (acoustic) waves, the density of the matter through which they run fluctuates, whereas, for instance, in the electromagnetic waves (light, radio, television, etc.), the intensities of electric and magnetic fields oscillate. There are temperature waves, waves of concentration in chemical reactions, epidemic waves, and so on and so forth. In a poetic way, one could say that waves penetrate the entire edifice of contemporary science.

The simplest possible type of waves are the monochromatic ones, when states change at each point according to a simple harmonic law, with a certain constant frequency (the sine or cosine law). The monochromatic sound waves are what we call the sound tones, or pitches. One can excite such waves using, for example, a tuning fork. The monochromatic light waves are generated by "their majesty" lasers. With a simple stick, while dipping it periodically up and down in water, one can make ripples that are pretty close to being monochromatic. A similar wave could be produced in our live chain on the road to St. Petersburg as well.

Imagine that each soldier not just simply raises his flag, but starts continuously and periodically waving it from side to side; and the next soldier follows the one before him with a certain fixed delay or *phase shift*. A wave takes off running along the chain. We bet our dear reader has seen something like that at a stadium during competitions, when the ardent (or simply bored) spectators start doing what is called "the wave".

These monochromatic events are quite pleasing to the eye, yet can they actually transmit information? – Obviously, no. Periodic oscillations do not tell us anything new, so no information is transmitted. Whereas, with the single toss of a hand the diligent servicemen did manage to communicate to St. Petersburg (more than $600\,km$ away from Moscow) the important news. What is the difference between these two kinds of waves? If we take an instant pictures of the motion for both situations, we will see that in the first case all the soldiers were involved in the motion, whereas in the second, only one of them. In other words, when a signal is being transmitted, the wave (whatever kind it was) at each moment is localized in space. One could imagine two, or three, or even several near-standing participants raise their hands at the same time. In such a case, the length of conveyed signal would increase. Having been able to generate signals of different length, one could send not only a message about a single event (such as the coronation having been accomplished, for example) but, in principle, any kind of information. Take, for instance, the famous Morse code (which was patented much later by the way, in 1854)[a].

Besides regiments of soldiers, there are, of course, other information-transmitting signals: light, sound, electric current, etc. It's interesting to note that any signal can be presented as a sum of the monochromatic waves with different frequencies. This possibility is ensured by the so-called *superposition principle,* stating that the oscillations of overlapping (interfering) waves at each point of a medium, should be just simply summed up. And, therefore, depending on the phase shift, oscillations can either amplify each other (for example, two identical waves with no phase shift will result in the oscillation of doubled amplitude, Figure 13.1, *a*) or extenuate each other (again: two identical waves being out of phase will cancel each other completely, Figure 13.1, *b*). It turns out also, that one can opt (or tune) the amplitudes and frequencies of combined monochromatic waves in such a manner that they will nullify each other in most places, except for a certain region where, conversely, they will amplify each other.

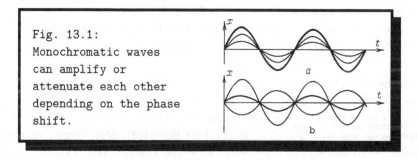

Fig. 13.1:
Monochromatic waves can amplify or attenuate each other depending on the phase shift.

The result of the summation of a large number N of waves with identical amplitude A_0, and frequencies lying within a small interval $2\,\Delta\omega$ around the basic frequency ω_0 is shown in Figure 13.2. It is like an instant photograph of a wave, showing variation of a fluctuating quantity A in different points of space at a fixed moment of time. There is a central maximum with $N\,A_0$ amplitude, and also numerous secondary peaks, though with quickly

[a]Surprisingly, the marine semaphore system where various positions of two (one red and one yellow) flags meaning different letters and figures appeared much later, in 1880. —A. A.

waning amplitudes, indicating that, indeed, the overlapping waves mostly extenuate themselves, having been noticeably amplified only in the vicinity of the central maximum.

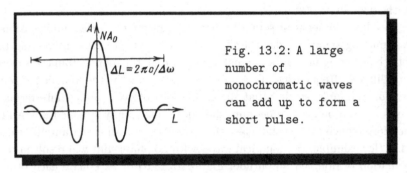

Fig. 13.2: A large number of monochromatic waves can add up to form a short pulse.

It is worth noticing that the central peak does not stand still, but moves with the wave's propagation speed. If the monochromatic component waves all move with the same velocity c (as in the case of electro-magnetic waves in a vacuum, for instance), the central maximum also moves with the same velocity c, keeping its constant width $\Delta L = \frac{2\pi c}{\Delta \omega}$. Hence, the time length of the running signal will be equal to $\Delta t = \frac{2\pi}{\Delta \omega}$.

Now one could easily write a simple yet quite fundamental relation:

$$\Delta \omega \cdot \Delta t \sim 2\pi.$$

So, the signal length and the width of the range where the frequencies of its components lie, are inversely proportional to each other. Qualitatively, such a relationship seems quite natural: if there is a segment of sinusoid, corresponding to a signal of rather long duration (Δt is large), then it must be an almost monochromatic wave and $\Delta \omega$ is small. However, if a short signal is required, it should be combined of many waves with different frequencies. Everyone, we believe, noticed glitches and noises generated on the radio on all bands when lightning strikes nearby.

Thus, each signal can be made of a set of monochromatic waves, or, just the same, each signal can be decomposed into such waves. The amplitude versus frequency dependence for monochromatic waves, composing a signal, is called the signal *spectrum*[b]. In this situation, for example, it will be a

---

[b]Although by spectrum physicists sometimes just mean the set of frequencies of monochromatic waves, making a signal, we stick to the more particular definition that takes into account the waves' amplitudes as well.

rectangle of altitude A_0 and width $2\,\Delta\omega$, depicted in Figure 13.3. This, of course, is a trivial spectrum; signal spectra, just like signals themselves, can have various, at times most peculiar shapes.

Fig. 13.3: The spectrum of the pulse depicted in Figure 13.2.

When we pronounce sounds, for example, we make the air vibrate in a certain way, so these vibrations propagate as sound signals of a certain shape. Their spectra strongly vary depending on whether we utter vowels or consonants. The vowels have spectra with two distinct characteristic peaks at certain frequencies (they are called *formants*). The spectra of consonants, on the other hand, are more "smeared", spread over the whole audible sound frequency range: Figure 13.4 represents the spectrum of the letter S sound. An entire method called *harmonic analysis* has been developed allowing us to find spectra of registered signals, as well as restore signals from their spectra.

Fig. 13.4: Spectrum of the consonant S sound.

Peculiar though it may sound, solid bodies are capable of "yelling" too. Thermal motion makes atoms in the crystal lattice oscillate, generating, thereby, elastic waves propagating inside the crystal. These oscillations are sound waves too. Their spectral maximum, however, lies in the area of extremely high frequencies – at the absolute temperature of about $5\,K$ it stays around $10^{12} - 10^{13}\,Hz$. In the audible frequency range, though, the amplitude of these oscillations is negligible; so that in order to hear "what the solids are talking about", one must use some special devices. And by listening to this chatting (studying the sound signal spectra, actually), researchers have already discovered plenty of very important secrets hidden in the solid-state.

But what kind of signals are used in practice to transmit information? For short distances sound signals have worked just fine, for as far back as the human history goes. The limitation though is that these types of waves tend to dissipate quickly. Yet, if one amplifies them at certain intervals (re-transmits them) such signals may travel quite a distance. In Africa, for example, until recently people were sending messages using the tam-tams, "drumming" a piece of information from one village to the next (similar to what the Russian soldiers did with their flags).

In the contemporary world most of the signals are transmitted in the form of electromagnetic waves, which can cover much longer distances, before dying out. For example, one can make an electromagnetic wave carry sound signals. In order to do that, the frequency of this electromagnetic wave (called the *carrier wave*) is kept constant, whereas its amplitude is varied (*modulated*) in accordance with the sound oscillation to be transmitted, Figure 13.5. This way a signal containing the required information is generated. Then, at the receiving end, the signal is "deciphered"– the envelope corresponding to the modulating sound signal is extracted. Such a sending–receiving method is called, therefore, *amplitude modulation* or *AM*. It's been employed in radio and television broadcasting[c].

There arises a new question though: how much information per unit of time can one actually transmit, with the help of waves? To sort this one out, let's look at the following situation. It's known that any number can be presented in binary notation, as a sequence of ones and zeros. In a

---

[c]Of course, after modulation, the electromagnetic wave no longer remains monochromatic. For example, in the case of simple amplitude modulation of the carrier wave of, say, frequency ω_0 and amplitude $A(t) = A_0 \left(1 + \alpha \sin \Omega\, t\right)$, in Figure 13.5:

$$x(t) = A(t) \sin \omega_0\, t = A_0 \sin \omega_0\, t + \frac{\alpha A_0}{2} \left[\cos(\omega_0 - \Omega)\, t - \cos(\omega_0 + \Omega)t\right].$$

You see that the spectrum even of this simplest modulation consists already of three different frequencies: $\omega_0 - \Omega$, ω_0 and $\omega_0 + \Omega$.

Fig. 13.5: Amplitude of amplitude-modulated carrier wave varies in accordance with the transmitted low-frequency signal.

similar way, any information can be written encoded into a row of successive pulses and pauses of certain duration. The signals can be transmitted by amplitude-modulated waves, for instance, Figure 13.6. The higher the desired speed of information transmission, the shorter these signals must be. Yet for reliably transporting the information the length of the signal shouldn't be shorter than the period of the carrier sinusoidal wave. This gives us right away the categorical limit on the maximum rate of information transmission. When one wants to raise the speed, one necessarily needs to increase the carrying frequency. Here, indeed, once again tells the considered relation for the time length of a signal: $\Delta t \approx \frac{2\pi}{\Delta\omega}$, where $\Delta\omega$ becomes comparable to ω_0.

Fig. 13.6: The simplest way to transmit digital information would be simply to chop the carrier into segments.

For example, to broadcast a musical program, it is sufficient to use electromagnetic waves with frequencies of around several hundred kilohertz ($1\,KHz = 10^3\,Hz$). The human ear is capable of sensing sound frequencies up to approximately $20\,KHz$, and, hence, the frequencies composing the signals, in this case, will be at least an order of magnitude less than that of the carrier wave. Yet, for transmitting television programs, this frequency range doesn't work. The images on the TV screen are generated 25 times per second, and they, in turn, consist of tens of thousands of separate dots (pixels). So the required modulation frequency is about $10^7 \, Hz$ and the corresponding carrying wave should have a frequency of several tens or hundreds megahertz ($1 \, MHz = 10^6 \, Hz$). That's why in TV technology, engineers had to utilize the *very-high (VHF)* and *ultra-high (UHF)* frequency bands[d]. Consequently, the *very-short* radio waves, with wave length of the order of one meter, are employed even though such waves can propagate only over rather limited (basically, within direct visibility) distances[e].

An even better candidate for fast information transmission would be ordinary light, having its frequency in the area of $10^{15} \, Hz$, which would boost the transmission speed by several tens of times. Although the idea is rather old (Alexander Bell[f] was actually the first who applied light signals for sending sound messages back in 1880), its technological feasibility arrived only recently with the development of high-quality sources of monochromatic light. Lasers, coupled with fiber-optical light guides, capable of transmitting light with extremely low propagation losses, as well as all the grandeur of modern electronics, allow efficient coding–decoding of these light signals.

Now we can with all certainty state that the age of copper wires is fading away and is being replaced with the coming epoch of ultra-fast information transmission networks based on fiber-optics technology.

---

[d]The terms VHF and UHF refer to the frequency (ν) bands $30-300 \, MHz$ and $300-3000 \, Mhz$ respectively. Sometimes both ranges are united into the single very-short wave band. The corresponding wave lengths are given by the usual formula $\lambda = \frac{c}{\nu}$, where $c = 3 \cdot 10^8 \, m/s$ is the speed of electromagnetic waves in vacuum. This gives $\sim 1 \, cm - 10 \, m$ for lengths of very-short waves. Actually in this band *frequency modulation (FM)* is used rather than AM. —A. A.

[e]Interesting enough the first television sets (with mechanical field scanning), made back in the 1920s, worked in the medium-wave (MW) band. The quality, due to the discussed problems, was very poor and the images were almost indiscernible. This made researchers and engineers switch to very-short waves and develop the electronic field scanning techniques. However, this MW television had its own advantage – because of the longer (compared to the very-short waves) propagating range, programs from Moscow, for example, could reach, say, Berlin without any TV satellites or retranslating stations.

[f]Whom you already know from Chapter 12.

Chapter 14

Why the electric power lines are droning

What is the Tacoma Narrows Bridge? The Tacoma Narrows Bridge was built in 1940. After Months of swaying up, down and side to side the Bridge collapsed. Taking with it the life of a poor dog (Tubby).

"Mr. H's World O'Physics."

Long ago, the Ancient Greeks noticed that tightly drawn string sometimes sounded melodically in the wind, as if it were singing. Perhaps, back then people knew the Aeolian harp, named so after Aeolus, the God of Winds in Greek mythology. It is made up of a frame (or an open box) with several strings stretched across, which is then placed where the wind can pass through it. Even a single string in such an instrument can generate quite a spectrum of different tones. Something of the same nature occurs, although with a far more limited tune variety, when the wind swings the cables of telegraph lines.

The puzzle of this phenomenon had long bewildered scientific pundits of the past, until the end of the seventeenth century when Sir Isaac Newton applied his newly developed analytical method to problems that we would call fluid dynamics.

According to the law, first stated by Newton, the resistance force acting on a body moving in liquid or gas is proportional to the square of the velocity v:

$$F = K \rho v^2 S.$$

Here S is the cross-sectional area of the body perpendicular to the direction of motion, ρ is the density of liquid (or gas), and K is just a proportionality coefficient.

Later it turned out that the formula does not apply universally. When the speed of the body is low compared to thermal velocities of molecules of the medium, the previous relation starts faltering. We have discussed already in Chapter 11 that for relatively slow-moving bodies the resistance force becomes linearly proportional to the speed (Stokes' law). This situation occurs, for example, when tiny droplets move in a rain cloud, or residue flakes precipitate onto the bottom of a glass, or the drops of substance $\mathcal{A}$ are roaming restlessly inside a Lava lamp (see the same chapter). However, in the modern world with its jet velocities, Newton's law of resistance force is much more common.

Either way, couldn't we find a satisfactory explanation to the phenomena of droning power lines or the tunes of the Aeolian harp just knowing these basic relations for the resistance? No, unfortunately, it is not at all that simple. Really, if the resistance force remained the same (or increased with the speed of the wind), the string would be simply drawn by the breeze without any oscillations at all.

So, where is the trick here? Well, it turns out that in order to understand the nature of the string vibration in this case, we can not get away with just a couple of rather general ideas not touching on flow mechanism. We ought to delve a bit deeper and discuss in more detail how liquid actually flows around a body at rest. (This is simpler than considering a body moving in the static liquid but does not affect the result, of course).

The case when the current is relatively slow is depicted in Figure 14.1. The liquid flow lines are passing smoothly around and behind a cylinder (the picture shows the cross section). Flow of this kind is called *laminar,* and the resistance force arises from the internal friction (viscosity) of the liquid and indeed is proportional to the velocity of the liquid (again – our reference frame is tied to the body). Both velocity and friction force at any point in such a flow are time-independent (the flow is *stationary*), and this rather insipid situation is of no interest to our Aeolian problem.

Now let's look at Figure 14.2. The flow velocity has increased and new whirling characters have appeared – the eddies or vortices, if you wish. Friction is not the definitive cause any more. Now it depends less on changes of momentum on a microscopic scale and more on scales comparable to the size of the body. The resistance force becomes proportional to the second

Fig. 14.1: Lines of slow laminar flow around long cylindrical wire.

Fig. 14.2: At higher flow velocities vortices appear behind the wire.

Fig. 14.3: In fast flows a periodic vortex trail is formed in the wake.

power of velocity, v^2.

Finally, see Figure 14.3, the flow velocity has grown even further, and the eddies are now aligned into neat well-ordered chains. Here indeed lies the answer to the riddle of the string's vibrations! These nicely structured tails of vortices break loose recurrently from the surface of the string and excite the string oscillations, like plucking fingers would do.

This arrangement of vortices tailing behind the body, was first discovered and studied experimentally at the beginning of the twentieth century. It found theoretical explanation in the works of the Hungarian scientist,

Theodore von Karman[a]. Now these periodic vortex wakes are known as the *Karman trails* (or even the Karman Vortex Street).

As velocity continues to increase even further, the vortices do not have enough time to spread over a large area of liquid. The "swirling" region narrows, the eddies start mixing with each other and the flow becomes chaotic, irregular *turbulent flow.* The latest experiments, though, show that for extremely high velocities there develops another kind of periodicity, but here we are getting well beyond the scope of our chapter, and the best we can do for the curious is to refer them to, say, the exciting book by James Gleick, titled *Chaos*.

It's worth noting that, although the phenomenon of Karman vortex trails may look like a rather academic example of just another nice quirk of nature, having not much practical significance, in reality it's quite the opposite. Electric power transmission lines, for example, swing in the wind when it's at a constant speed because of these periodically born and released vortices. And this inevitably creates quite powerful stress at the wire fastenings to the supporting towers, which, if neglected, could (and, unfortunately, have) led to their breakage (sometimes very dangerous). The same holds for tall industrial smoke stacks.

Probably the most renowned engineering calamity of this kind happened in Tacoma (Washington), in November 1940, when a new auto bridge (two lanes, about half a mile long) built just a couple of months before, Figure 14.4, started swaying violently and collapsed, luckily not killing or

Fig. 14.4: Growing oscillations excited by turbulent vortices led to the collapse of Tacoma Narrows Bridge.

[a]T. von Karman, (1881–1963). Having greatly contributed to the field of fluid dynamics, he is known as the father of supersonic flight. Hungarian-born, he worked for the US government during World War II, and then for the NATO Aeronautical Research and Development Advisory Group.

injuring anybody (except for the legendary dog mentioned in the epigraph).

A special committee from the Federal Works Agency was appointed to investigate the accident, with T. von Karman as one of its members, by the way. And the committee's conclusion read that the Tacoma Narrows Bridge had crashed due to "forced vibrations excited by random action of turbulent wind". Well, a new bridge was built after a while, but this time with a completely different profile for the wind-exposed surfaces, eliminating, thereby, the cause of unruly vibrations.

Chapter 15

The footprints on the sand

You walked with me, footprints in the sand,
And helped me understand where I'm going.

Simon Cowell, et al., *"Footprints on the Sand."*

Have you ever thought whether, when walking on a beach, you compress the sand under your feet? On the face of it, stepping onto sand presses the grains together. But actually things may turn out quite the opposite way. Here is the proof – the footprints left on wet sand stay dry quite a while. The English scientist known for his works on hydrodynamics, O. Reynolds[a], noted in his talk at the meeting of the British Association in 1885 that when the foot steps onto sand still wet after the ebb, the surrounding area immediately becomes dry... According to him the pressure of the foot loosens the surrounding sand and the stronger the pressure the more water is absorbed. This makes the sand dry until enough water comes from below.

But why does pressure widen intergranular spaces so that the available water no longer fills them? For scientists of the nineteenth century this was not an idle question. The answer was immediately related to the atomic structure of matter. This is the topic of the present chapter.

---

[a]O. Reynolds, (1842–1912), English physicist and engineer, specialist in theory of turbulence, theory of viscous flows, and theory of lubrication.

15.1   The dense packing of balls

Is it possible to fill an entire space with rigid balls of the same radius? Of course not; there will always be voids in between[b]. The fraction of space occupied by the balls is called the *packing density*. The closer the balls lie, the less space is left between them, and the higher the packing density. But when does the density reach the maximum? The answer to this question gives a clue to the mystery of the footprints on the beach.

Let us start from a simpler case and study the packing of equal circles in a plane. Dense packing of circles can be obtained by inscribing those into the cells of a mosaic *(tilings of the plane)* composed of equal regular polygons. There are only three variants: to cover the plane by equilateral triangles, by squares, and by hexagons. The packing of circles in the cells of the square and hexagonal mosaics is shown in Figure 15.1. It is easily seen that the second pattern *(b)* is more economical. The accurate calculation (that you may carry out yourselves) proves that in this case 90.7 percent of the surface is covered by circles whereas for squares, *(a)*, the portion makes only about 78 percent. The hexagonal packing is the densest possible in the plane (or, as modern physicists like to say, in two dimensions). Maybe this is the reason that bees use it for the honeycomb.

A dense spatial packing of balls may be realized as follows. Let us first prepare a dense plane layer of balls placed on a flat surface in the economical order described above. We shall call it the X-layer. Now try to put a similar hexagonal second layer on top of it. We could do this so that each of the balls of the upper layer would lie right over a ball of the lower one, as if we were filling cells of invisible honeycomb. However in this XX-packing too much empty space will be left. The volume occupied by balls when laid in this manner is only 52 percent of the whole.

To augment the density one must simply put the upper balls into the holes formed by three touching balls below. (This may be called the XY-arrangement). But it is impossible to fill all the holes at once – one of the two adjacent holes will always remain free, Figure 15.2. Therefore when putting on the third layer we face a choice. We may either put balls above the holes of the ground X-layer which were left free by the second Y-one

---

[b]In principle, it is possible to fill the space with balls provided that their radii $r_1, r_2, \ldots$ form an infinite sequence and $\lim_{n \to \infty} r_n = 0$. However this is of little interest for solid-state physics. —A. A.

Fig. 15.1: Planar packing of circles inscribed
into the cells of tilings of the plane by equal
regular polygons.

(one of these points is denoted by A in Figure 15.2, b) and build a new
Z-layer; or we can place them right over the balls of the base (the point
B in Figure 15.2, b) in the X-order. Regular spatial patterns are obtained
if successive layers follow periodically one of these prescriptions, that is,
either $XYZXYZ\ldots$ or $XYXY\ldots$ As a result we obtain the two ways of
spatially packing balls depicted in Figure 15.3. In both cases the balls fill
about 74 percent of the space.

It is easy to count that every ball in these packings touches 12 others
and the points of contact are vertices of a 14-faced polyhedron[c]. The faces
of these polyhedra are alternating squares and equilateral triangles. Say,
the first choice (Figure 15.3, a) produces the cuboctahedron[d] shown in
Figure 15.4.

So far we have only studied how to arrange balls so that they fitted
a periodical spatial honeycomb[e]. But is it possible to construct a dense

---

[c]The Greek word *tettarakaidecahedron* ($\tau\epsilon\tau\tau\alpha\rho\alpha\kappa\alpha\iota\delta\epsilon\kappa\alpha\epsilon\delta\rho\upsilon\nu$) is out of common use.
—A. A.

[d]Cuboctahedron belongs to the so-called Archimedean solids. This class comprises 13
convex polyhedra with congruent vertices and surfaces composed of regular polygons of
two different types. The name *cuboctahedron* itself belongs to J. Kepler (see page 128).
The packing in Figure 15.3, b does not correspond to any of Archimedean bodies since
it generates two different types of vertices. —A. A.

[e]Another way to say this is that the centers of the balls formed a periodic lattice.—A. A.

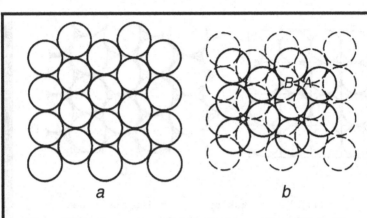

Fig. 15.2: Compact spatial packing of equal balls.
(Dashed lines show the lower layer.)

Fig. 15.3: In three dimensions there are two
compact arrangements of identical balls.

packing without this condition? An example is presented in Figure 15.5.
The balls in planes mark out the sides of concentric regular pentagons. The
nearest balls of the same pentagon are in contact, but pentagons within a
layer are separated. The sides of pentagons in alternating layers contain
even and odd numbers of balls respectively. The packing density of this
arrangement is 72 percent which is not much less than that of the dense

Fig. 15.4: Kepler's
cuboctahedron.

hexagonal arrangements shown in Figure 15.3. There is a way to pack balls
so that the centers do not form a lattice and the packing density reaches
74 percent. Yet the question whether denser packings exist remains open.

Fig. 15.5: Pentagonal
packing of balls.

Let us return to the footprints on the beach. Now we know that there
are special arrangements of balls that leave only a small amount of void in
between. If one disturbs such an arrangement displacing, for instance, the
balls of one layer away from the holes of the next one the voids will grow.
Certainly nobody cares to arrange the grains of sand in a special way. But
how could we force sand to be densely packed?

Remember common sense. What do you do when pouring grain into a
can? You shake the can gently in order to fill it better. Even if the grain
is in clumps patting helps it fit in.

The scientific investigation of this trick was undertaken in the 1950s by
the British scientist G. Scott who loaded spherical flasks of different radii
with ball bearings. When they were filled without shaking, so that the balls

found places randomly, the empirical dependence of packing density on the number of balls had the form:

$$\rho_1 = 0.6 - \frac{0.37}{\sqrt[3]{N}},$$

where N was the number of balls. You can see that if the number of balls is very big (it had reached several thousands in the experiments) then the density tends to become constant and corresponds to filling 60 percent of the space. But shaking the container while filling helps to increase the density:

$$\rho_2 = 0.64 - \frac{0.33}{\sqrt[3]{N}}.$$

Yet even in this case, it is much less than 74 percent which corresponded to the regular packing of balls.

The results of the experiment deserve attention. Why is the addition inversely proportional to $\sqrt[3]{N}$? The balls near the walls of the flask are in a special position with respect to those inside. This affects the density of packing. The magnitude of their contribution is proportional to the ratio of the surface ($\sim R^2$) to the volume ($\sim R^3$) of the container being inversely proportional to the size of the system (R). By the volume of the system we mean the entire volume occupied by the balls including spaces between them. The size of the system is $R \sim \sqrt[3]{N}$ since the volume is proportional to the number of balls. Dependencies of this type often appear in physics when one has to take into account surface effects.

You see that accurate experiments are in agreement with common sense and prove that shaking granular substances helps to enlarge the density of packing. But what is the reason? Remember that positions of stable equilibrium always correspond to minima of potential energy. A ball may forever lie steadily in a hole but it will immediately roll down from a bump. Something of the sort happens here as well. Shaking the flask makes the balls roll to free spaces so that the density of packing is increased and the total volume of the system becomes smaller. As a result the level of balls in the container goes down. Consequently, the center of mass of the system gets lower and the potential energy is decreased.

Now, at last, we can work out clearly enough what happens to the wet sand. Incessant surf agitates it until a dense packing of grains is formed. By stepping onto the sand with your foot, you disturb the arrangement of granules and augment intergranular pores[f]. Water from the upper sand layer soaks down to fill the pores. This looks like the sand is drying. Taking the foot away restores the dense packing and the depression left by the foot gets filled with water expelled from narrowing voids. However sometimes after strong compression the dense packing can not be recovered. Then the footprint will become wet only as water rises from below and fills the widened pores.

It is interesting that these features of granular matter were well known to Indian fakirs. One of their tricks consisted of sticking a long thin dagger several times into a narrow-necked vase poured with rice. At some point the dagger gets stuck in the rice and it is possible to lift the vase holding it by the dagger handle.

Evidently the secret was that piercing randomly poured rice helped to "optimize" the packing of grains just like shaking would. One may imagine this like a sort of compression wave propagating in a loose medium. In the beginning the grain was packed compactly right around the blade but lay freely in the bulk and near the walls. The "front" of the wave (surely a rather smooth one) divided the dense core from the loose environment. The front advanced further with every stab and when at last it reached the walls of the vessel the rice throughout the volume became densely packed. In other words the possibilities to compress it further were exhausted. The properties of the substance changed drastically, it became "incompressible". And this was the moment when the dagger gets stuck, since the pressure of grains on the blade and consequently the friction are enough to hold it[g].

⚠ **Caution!** In case you've decided to surprise your party fellows, please, avoid using glass flasks and china vases. The result can be quite unexpected.

15.2 The long-range and short-range orders

Of course the atoms of which all bodies are built are not rigid balls. Yet simple geometrical arguments help us to understand the structure of matter.

The first to apply a geometrical approach was the German scientist

---

[f]Note that according to Reynolds (see page 121) this refers to the sand *around* the footprint whereas right under the sole it stays densely packed. —A. A.

[g]The paragraph was added to the English edition by —A. A.

Johann Kepler[h] who in 1611 put forward the idea that the hexagonal form of snowflakes is related to the dense packing of balls. Mikhail Lomonosov[i] in 1760 first delineated the most compact cubic packing of balls and used that to explain the forms of crystal polyhedra. The French abbot R.-J. Haüy[j] noticed in 1783 that all crystals may be constructed of a plethora of repeated parts, Figure 15.6. He explained the regular form of crystals by suggesting that they are built of identical little "bricks". Finally in 1824 the German scientist A. L. Seeber proposed the model of crystal composed of regularly set little spheres interacting like atoms. The dense packing of the spheres corresponded to the minimum of potential energy.

Fig. 15.6: Pictures from the atlas by R.-J. Haüy published at the start of the nineteenth century.

[h] J. Kepler, (1571–1630), German astronomer, creator of celestial mechanics. The famous Kepler laws of planetary motion laid the base for Newton's discovery of the law of gravitation. Kepler's interest in polyhedra was a tribute to the idea of the world ruled by mathematical harmony. According to Kepler ratios of the radii of planetary orbits in the Solar system could be related to properties of regular and uniform polyhedra.

[i] M. V. Lomonosov, (1711–1765), the first scientist of world importance in Russian history. Successfully worked in natural sciences including physics, chemistry, material science, as well as in literature, poetry, and painting. Founded the Moscow University.

[j] R.-J. Haüy, (1743–1822), French crystallographer and mineralogist.

Structures of crystals are the subject matter of a special science called *crystallography*. Presently, the periodic arrangement of atoms in crystals is a well-established fact. Electronic microscopes offer the chance to see this. The tendency to positive close packing does exist in the atomic world. About 35 chemical elements crystallize so that their atoms are situated like the balls depicted in Figure 15.3. The centers of atoms (or, to be precise, atomic nuclei) make up in space a so-called *crystal lattice* that consists of periodically repeated units. The elementary lattices that may be constructed by periodically shifting a single atom are called *Bravais lattices* (after the French naval officer Auguste Bravais[k] who was the first to develop the theory of spatial lattices).

There are not many Bravais lattices – only 14 of them. The reason is that most symmetrical elements don't survive in periodic lattices. For instance, a regular pentagon may be turned around the axis passing through the center and it will coincide with the original five times per revolution. One says that it has a fivefold symmetry axis. However a Bravais lattice can not have a fivefold axis. If such a lattice existed its nodes would be vertices of regular pentagons and those, in turn, would cover the entire plane without breaks. But it is well known that there is no tiling of the plane by pentagons, Figure 15.7!

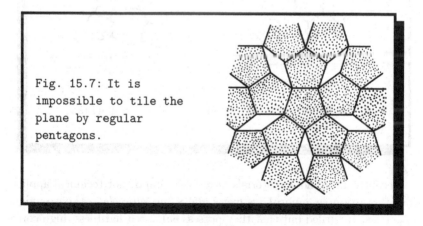

Fig. 15.7: It is impossible to tile the plane by regular pentagons.

[k]A. Bravais, (1811–1863), French crystallographer.

So all crystals can be composed of repeated units. This property is called *translational symmetry*. One may also say that there is a *long-range order* in crystals. This is the main property that distinguishes crystals from other bodies.

There is, though, an important class of substances which are deprived of the long-range order. These are *amorphous* substances. Liquids are an example of the amorphous state. But solid matter also may be amorphous. The structure of glass is portrayed in Figure 15.8 together with that of quartz which has the same chemical composition. But quartz is a crystalline substance in distinction from the amorphous glass. The absence of a long-range order does not mean that atoms in glass are situated chaotically. You may see in the picture that a certain ordering of the nearest neighbors is preserved even in glass. This is called a *short-range order*.

Fig. 15.8: Structures of quartz, *(a)*, and glass, *(b)*.

Recently amorphous materials have found important technical applications. For example, amorphous metal alloys *(metallic glasses)* have unique properties. It turned out that they possess enhanced hardness, high corrosion resistance, and exhibit an optimal compromise of electric and magnetic characteristics. Metallic glasses are obtained by means of an extremely fast cooling of liquid metal: the velocity of which must be of the order of several thousands degrees per second. This may be realized by sputtering tiny droplets of metal onto the surface of a rapidly spinning cold disk. The droplets get squashed against the disk forming a film of several microns thickness and the instantaneous removal of heat simply leaves no time for the atoms to arrange properly when cooling[l].

Interesting studies that shed light on the structure of amorphous solids were carried out in 1959 by the English scientist J. Bernal[m]. Equally sized balls of plasticine were randomly put together and compressed into a big lump. The polyhedra obtained after disjoining them turned out to have mainly pentagonal faces. The experiment was repeated with lead pellets. If the pellets had been laid densely and regularly then the compression remolded them into almost regular rhombododecahedrons[n]. On the other hand, if the pellets had been poured unintentionally they transformed into irregular 14-faced polyhedra. Among the faces of those were tetragons, pentagons and hexagons forms, but pentagons prevailed.

In modern technology it is often necessary to densely pack the elements of a contrivance. For instance, Figure 15.9 shows a cross section of a superconducting cable made of a large number of superconducting wires enclosed in a copper envelope. At first the wires were cylindrical but after rolling they became hexagonal prisms. The more densely and accurately the wires were packed, the more regular hexagons were at the section. This is evidence of the high quality of the cable. If the density of packing is interrupted, pentagons appear at the section.

Fivefold symmetry is widespread in nature. Figure 15.10 presents a photograph of a viral colony. What a striking similarity it has with the pentagonal packing of balls portrayed in Figure 15.5! Paleontologists even

---

[l]In the method of melt spinning, a jet of molten metal is propelled against the moving surface of a cold, rotating copper drum. A solid film of metallic glass is spun off as a continuous ribbon at a speed that can exceed a kilometer per minute.

[m]J. Bernal, (1901–1971), English physicist, specialist in X-ray diffraction analysis, studied structures of metals, proteins, viruses, etc.

[n]Rhombododecahedron (or rhombic dodecahedron) is a polyhedron with 12 rhombic faces and 14 vertices. It may be obtained as a result of uniform squeezing of the hexagonal packing depicted in Figure 15.3, *a*. A way to construct it geometrically is to draw mutual tangential planes of a ball and all its neighbors. In the compact packing shown in Figure 15.3, *a* it plays the role of the unit cell which belongs to hexagons in the planar tiling, Figure 15.1, *b*. Each ball is inscribed into a rhombododecahedron and those fill the entire space. —A. A.

Fig. 15.9: Cross
section of
high-quality
superconducting cable.
After rolling
cylindrical wires
became hexagonal.

use the presence of fivefold axes in fossils as a proof of their biological
(contrary to geological) origin. See how far away from the deserted beach
the footprints have led.

Fig. 15.10:
Electron-microscope
photograph reveals the
fivefold symmetry of
viral colony.

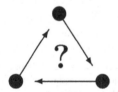

Was it important that Indian fakirs used metal
vases with long slim necks for their tricks? What
ratio of the volumes of the neck to the body of the
vase must be used?

Chapter 16

How to prevent snowdrifts

Sections of roads and railroad tracks passing through hollows are often covered with snow even if there have been no snowfalls recently. Why does this happen? Certainly, the answer is simple: the snow has been brought by wind. However it took a good deal of investigation to understand in detail the mechanism of the process.

In 1936 the English geologist Ralph Bagnold[a] studied wind transport of sand in an air tunnel. He discovered that unless the wind velocity is larger than some critical value v_1 the sand would not move. If the velocity of air flow is higher than v_1, but yet less than another value, v_2, the mass of sand may stay at rest. However, an occasional sand grain falling from above brings several rebound particles. These get caught by the wind and, when falling down, knock out more sand grains from the base layer. As a result the sand gets carried by the wind in a kind of leaping motion. In case that velocity exceeds v_2 the wind lifts and blows along substantial clouds of sand. The density of the clouds decreases with height. Trajectories of sand grains may be viewed in Figure 16.1.

Now we can explain why the wind fills dips with snow. Look at the picture of flow lines in Figure 16.2. It is obvious that when traversing a hollow the air flow widens and its velocity lessens. This disturbs the balance between deposited and lifted particles: more particles fall down than are taken away. So the depression gets gradually filled with snow.

Analogous processes take place when the snow carried by wind encounters an obstacle, say, a tree. Meeting the trunk the incident flow turns up

---

[a]R. A. Bagnold, (1896–1990), English geologist, expert in the mechanics of sediment transport and eolian (wind-effect) processes.

Fig. 16.1:
Trajectories of sand
particles in air flows
of different strength.

Fig. 16.2: Flow lines
rarefy when wind
crosses a hollow.

and the ascending current of air is formed. This current digs a deep hole on the windward side of the tree. At the same time in front of the hole and behind the tree the speed of wind is smaller and a mound is heaped up.

This phenomenon helps prevent snow-binding in low-lying road sections. A protecting fence is made of wooden planks at a certain distance windward of the road. Behind the fence a lee zone of steady light wind is established where all the snow precipitates.

The same mechanism explains the motion of sand dunes. A strong enough wind blowing against a dune picks up sand on the windward side. On the rear of the dune the air flow slows down and the sand falls back down. As a result, with time, dunes get moved along inch-by-inch with the wind and the dunes "wander".

Chapter 17

The incident on the train

Not so long ago the authors of this book had to return from Venice to Naples by express train. The train ran fast (its velocity was about $150\,km/h$) and landscapes that looked like paintings by Renaissance masters flitted by outside the window. In perfect accord with their canvases, the country was hilly, and we sometimes flew over a bridge or dove into a tunnel.

In one especially long tunnel between Bologna and Florence, we suddenly felt a dull pain in our ears, as happens to air passengers when taking off or landing. It was clear that the same sensation visited all our fellow travelers, who swayed their heads trying to get rid of the unpleasant feeling.

But when the train finally burst out from the narrow tunnel the discomfort passed. However one of us, who wasn't used to such surprises on the railways, got interested in the origin of this phenomenon. Since it was evidently connected with a pressure jump, we started a lively discussion of the possible physical causes.

At first glance it seemed to us that the air pressure in the gap between the tunnel walls and the train had increased in comparison with the atmospheric one, but this assumption became less and less obvious as the discussion went on. In such matters mathematics is the best judge, so we attempted to approach the problem quantitatively. Soon we arrived at an explanation which follows.

Let's consider a train with a cross-sectional area S_t that moves at a velocity v_t in a long tunnel with a cross-sectional area S_0. First of all, let's switch to the inertial coordinate frame associated with the train. We'll take the air flow as stationary and laminar and ignore its viscosity. The motion of the tunnel walls relative to the train need not be taken into account in this case – because of the absence of viscosity it doesn't influence the air flow. We'll also assume that the train is sufficiently long so that one could ignore end effects near the front and rear cars. We shall assume that the air pressure in the tunnel is constant and does not vary along the whole train.

You see, that gradually eliminating minor details, we'd passed from the actual movement of the train to a simplified physical model that could be analyzed mathematically. Here it is.

We have a long tube (the former tunnel) with air being blown through it and a cylinder with streamlined ends (the former train) coaxially fixed inside[a], Figure 17.1. Far away from the train (at the cross section $A - A$) the air pressure equals the atmospheric one p_0. Velocity of the air flow at this section is equal in magnitude and opposite in direction to the velocity of the train $\vec{v_t}$ with respect to the ground. Let's examine some cross section $B - B$ (just in case, we may place $B - B$ far enough from the ends of the train so that our assumptions are justified). We'll denote the air pressure in this cross section by p and the air velocity by v. These values can be linked with v_t, and p_0 by means of the Bernoulli[b] equation:

$$p + \frac{\rho v^2}{2} = p_0 + \frac{\rho v_t^2}{2}, \qquad (17.1)$$

where ρ is the air density. Equation (17.1) contains two unknowns, p, and v; so in order to determine p we need one more relation. This is provided by the condition of conservation of the air flow. According to this the mass of air passing through any cross section of the tube in a unit of time is constant and equals:

$$\rho v_t S_0 = \rho v (S_0 - S_t). \qquad (17.2)$$

This equation expresses the fact that the air mass can neither appear nor disappear while it flows through the tube. It's usually called the condition of flow continuity.

As you probably noticed, we took the air density in equations (17.1) and (17.2) to be constant. For this assumption to be valid, two conditions

---

[a]Note that actually we replaced the ordinary railway tunnel with an air tunnel like those where airplanes are tested.

[b]Daniel Bernoulli, (1700–1782), Swiss physicist and mathematician born in the Netherlands (son of the Swiss mathematician Johann Bernoulli); formulated the fundamentals of theoretical hydrodynamics.

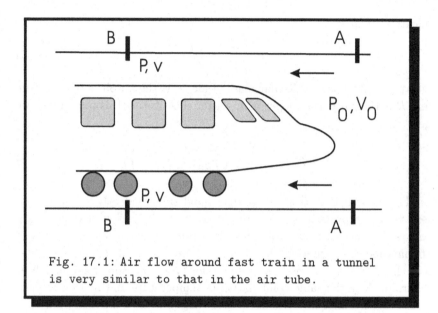

Fig. 17.1: Air flow around fast train in a tunnel is very similar to that in the air tube.

must hold. The first one requires that the pressure jump we are looking for, Δp, must be much less than the pressure itself: $\Delta p \ll p$. If the air temperature does not change along the tube, then, according to equation (17.4), its density is proportional to pressure: $\rho \propto p$. For small Δp we may neglect the density change $\Delta \rho = \rho \frac{\Delta p}{p} \ll \rho$. We shall see later that this is indeed the fact. The second condition concerns flow velocities at different sections of the tunnel. In order that the density is uniform throughout the tube there must be enough time for air to come to equilibrium. This means that velocities of flow must be much less than the root-mean-square velocity of chaotic thermal motion of the molecules. It's just this velocity that determines the characteristic time required to establish the constant equilibrium gas density on the macroscopic scale.

Eliminating the velocity v from equation (17.1) by means of equation (17.2), we get:

$$p = p_0 - \frac{\rho v_t^2}{2} \left(\frac{S_0^2}{(S_0 - S_t)^2} - 1 \right). \tag{17.3}$$

The air density ρ can be expressed in terms of p by the Clapeyron–Mendeleev equation (as we have already done in Chapter 12, see equation (20.3)):

$$\rho = \frac{m}{V} = \frac{p_0 \, \mu}{R \, T}, \qquad (17.4)$$

where $\mu = 29$ is the molecular mass of air, T is the absolute temperature and R is the gas constant per mole. After substituting this into (17.3), we get:

$$p = p_0 \left[1 - \frac{\mu \, v_t^2}{2R T} \left(\frac{S_0^2}{(S_0 - S_t)^2} - 1 \right) \right]. \qquad (17.5)$$

The combination $\frac{\mu \, v_t^2}{2R T}$ in the right-hand side, evidently, is dimensionless. So the expression $\sqrt{R T / \mu}$ must have the dimension of velocity. Up to a coefficient it's easy to recognize in it the root-mean-square velocity of thermal molecular motion. But in aerodynamical problems another physical characteristic of gas, the sound velocity v_s, is more relevant. It is determined by the same combination of temperature and molecular mass as the root-mean-square velocity of molecules, but the numerical value of v_s includes in addition the so-called adiabatic index γ. The latter is a number of the order of unity characteristic of a gas (for air, $\gamma = 1.41$):

$$v_s = \sqrt{\gamma \frac{R T}{\mu}} \, . \qquad (17.6)$$

Under normal conditions, $v_s = 1\,200 \, km/h$. With the help of equation (17.6) we can bring the expression (17.5) to the form that will be convenient for the further discussion:

$$p = p_0 \left[1 - \frac{\gamma}{2} \frac{v_t^2}{v_s^2} \left(\frac{S_0^2}{(S_0 - S_t)^2} - 1 \right) \right]. \qquad (17.7)$$

Now it's time to stop and think a little. We calculated the pressure along the skin of the train inside the tunnel. But our ears ached not because of the pressure itself but because it had changed in comparison with the pressure p' that was there when moving in the open[c]. We can easily determine this

---

[c] Here we should point out two circumstances. First, in biophysics there is the so-called Weber–Fechner law. According to this law any changes in the environment can be detected by organs of sense only if the relative change of parameters exceeds some threshold value. The second is that in a long tunnel our organism adapts to the new conditions and the discomfort disappears. Nevertheless, it comes back at the exit.

pressure directly from equation (17.7), noticing that the open air can be considered as a tunnel of the infinite cross section $S_0 \to \infty$. So we have:

$$p' = p_0.$$

This result was sufficiently evident without calculation, though.

It's interesting to observe that the relative pressure difference is negative:

$$\frac{\Delta p}{p_0} = \frac{p - p_0}{p_0} = -\frac{\gamma}{2} \left(\frac{v_t}{v_s} \right)^2 \left(\frac{S_0^2}{(S_0 - S_t)^2} - 1 \right). \qquad (17.8)$$

From this we can see that when a train enters a tunnel the pressure around it decreases, contrary to what we might expect at first. Now let's estimate the magnitude of the effect. As we have mentioned before, $v_t = 150\,km/h$ and $v_s = 1\,200\,km/h$. For narrow railroad tunnels the ratio $S_t\,/\,S_0$ is about $1/4$ (for there were two tracks in our tunnel). So we find that:

$$\frac{\Delta p}{p_0} = -\frac{1.41}{2} \left(\frac{1}{8} \right)^2 \left(\left(\frac{4}{3} \right)^2 - 1 \right) \approx 1\,\%.$$

This value may seem pretty small, but if we take into account that $p_0 = 10^5\,N\,/\,m^2$ and take the area of the eardrum to be $\sigma \sim 1\,cm^2$, we get an excess force $\Delta F = \Delta p_0 \cdot \sigma \sim 0.1\,N$, which may turn out to be quite noticeable.

So it seemed that the effect was explained, and we could call it quits. But something still worried us about the final equation. Namely, from expression (17.8) it followed that even in the case of an ordinary train moving with normal velocity, so that[d] $\frac{v_t}{v_s} \ll 1$, the value of $|\Delta p|$ might reach and even exceed the normal pressure p_0 in sufficiently narrow tunnels! Clearly, within the framework of our assumptions we were getting the absurd result that the pressure between the walls of narrow tunnel and train became negative!

Wait a minute! We probably missed something that hampers the validity of our formula. Let's take a closer look at our findings. If $\Delta p \sim p$, then

$$\frac{v_t}{v_s} \left(\frac{S_0}{S_0 - S_t} \right) \sim 1,$$

---

[d]The omnipresent ratio of velocities $M = v\,/\,v_s$ in aerodynamics is called the Mach number after the Austrian physicist Ernest Mach, (1838–1916).

and, consequently,

$$v_t \, S_0 \sim v_s \, (S_0 - S_t).$$

Comparing the last equation with the continuity equation (17.2), we begin to understand the situation. If Δp becomes of the order of p_0 the velocity of air flowing in the gap between the train and the walls of this narrow tunnel turns out to be of the order of the speed of sound. That is, we can't speak of laminar air flow any longer and the previously smooth flow becomes turbulent.

So the correct condition for the use of equation (17.8) is not merely $v_t < v_s$ but the more rigid one:

$$v_t \ll v_s \left(\frac{S_0 - S_t}{S_0} \right).$$

It's evident that for real trains and tunnels this condition is always met. Nevertheless, our investigation of the limits of applicability of equation (17.8) wasn't just an empty mathematical exercise. Physicists must always understand the limits of validity of their results. Besides there is a quite practical reason to take it seriously. In the last few decades fundamentally new forms of transportation, including high-speed trains, were discussed more frequently. One of the projects exploited a magnetic cushion produced by a powerful superconducting magnet. By the early 1990s a prototype *maglev* (an abbreviation for *magnetic levitation*) train in Japan could carry 20 passengers along the 7 km test track at a maximum speed of 516 km/h – almost half of sonic speed! The vehicle hovered above metal rails supported by a strong magnetic field and resistance to its motion was determined solely by aerodynamic effects.

The next step in developing this transport was the idea of, believe it or not, enclosing the train in a hermetically sealed tube and reducing the aerodynamical factor by pumping the air out! You see how close this problem is to the one that has captivated us. But here physicists and engineers have encountered a much more complex case of $v_t \sim v_s$ and $S_0 - S_t \ll S_0$. Therefore the air flow is not laminar, and the temperature of the air changes considerably along the train.

Modern science doesn't have ready answers to the questions which appear when solving these problems. But even our simple estimate allows, in principle, an understanding of when new effects come into play and become important.

By the way, here are some more physical questions that might pop up during a train ride.

(1) Why does the noise from a moving train increase considerably when the train enters a tunnel?
(2) Which of the two rails of a track is worn down faster in the Northern Hemisphere? And what about the Southern Hemisphere?

Why do express trains on close parallel tracks slow down when meeting?

PART 3

Physics in the kitchen

In this part of the book we will explain how the laws of physics govern preparation of different dishes and drinks, starting from everyday tea and coffee and finishing with delicious wines and champagnes.

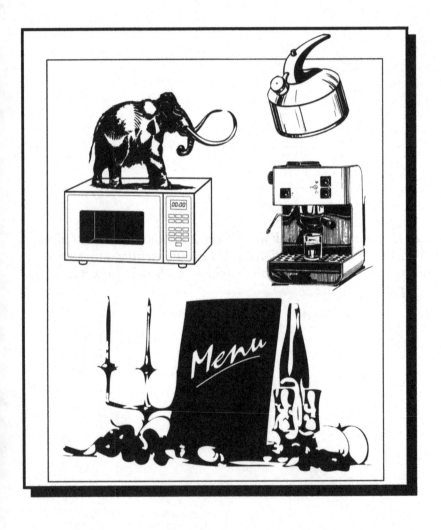

Chapter 18

Ab(out) ovo

For the Words are these:
    That all true Believers shall break their Eggs at the convenient End:
    and which is the convenient End, seems, in my humble Opinion, to be left to every Man's conscience, or at least in the power of the Chief Magistrate to determine.

Jonathan Swift, *"Gulliver's Travels."*

Literally, the Latin expression *ab ovo* means "from the egg". Figuratively, it indicates the earliest possible stage, the very origin of something. The ancients used it in order to emphasize the fundamental importance of an event. Thus, without noticing, they resolved the famous paradox "which came the first, the chicken or the egg?" in favor of the egg. But we shall leave this perennial dispute aside[a] and look instead at various physical phenomena around this seemingly unexciting object, the chicken's egg.

As everyone remembers from the school reading of *Gulliver's Travels*, the two great empires of Blefuscu and Lilliput were engaged in a most obstinate war that started after the Emperor of Lilliput published an edict commanding all his subjects, upon great penalties, to break the smaller end of their eggs. Gulliver believed (and was in agreement with the doctrine of the great Lilliputian prophet Lustrog) that it was really a private matter which particular end of an egg to break. The authors respect their opinion but, for the sake of curiosity, it would be interesting to find out which end

---

[a]For those interested in paradoxes, we would recommend the excellent collection *Paradoxicon* by Nicholas Falletta.

of the egg shell is it *easier* to break. The solution would give one a clear
winning strategy in the fierce battles of "hard-boiled eggs" that often break
out at the Easter table.

What is the right tactic: to attack the opponent or to grant him the
first strike? Which egg to take: a big or a little one? To hit the small end
or the big one? These are the main strategic questions of such a fight. The
common opinion is that the attacking side gains the advantage. However, if
both eggs move uniformly it makes no difference which is moving and which
is static. One may see that without breaking the eggs. Recall Galileo's
relativity principle and look at the problem from the moving reference frame
where the attacker is at rest. In this frame he automatically changes from
the "aggressor" to the suffering side.

Now, let us look closer at the actual collision of the two eggs. Let the
eggs be exactly alike, i.e. suppose that their size, their shape, and the
strength of their shells (the breaking stress σ_b) are the same. The eggs
collide along the common axis, one with its small end and the other with
the big end at the point of collision (Figure 18.1).

Fig. 18.1: Two eggs
colliding along the
common axis.

According to Newton's third law, the forces $\mathbf{F}_1$ and $\mathbf{F}_2$ acting on the
colliding eggs are equal and oppositely directed. Let's represent $\mathbf{F}_1$ and
$\mathbf{F}_2$ as sums of distributed forces $\mathbf{f}_1$, $\mathbf{f}_1'$ and $\mathbf{f}_2$, $\mathbf{f}_2'$, directed along the corresponding tangents to the colliding surfaces[b]. The magnitude of stress
f depends on the angle between the tangents: $f = \frac{F}{2\cos\alpha/2}$. Therefore,
the greater the angle, the greater the "cracking" forces acting on the shell
during the collision. The angle is defined by the radius of curvature of the
surface. The picture clearly proves that it is indeed advantageous to strike

---

[b]Note that due to the elasticity of the shells, the colliding eggs touch each other at a
tiny but finite spot.

the opponent's egg with the small end of your egg. Another reason in favor of this strategy is that the air cell in the large end weakens the wide side of the egg even more (try to explain this).

Note that the previous analysis suggests a practical ruse: even if the experienced opponent turns his egg with its small end forward, you can heighten your chances by attacking him and hitting his egg slightly to the side of the tip, rather than straight on. There the curvature of the shell is less and therefore the stress produced by deformation increases[c].

In toy and souvenir stores one can find a funny toy called a *"tippee top"*. It is shaped like a truncated ball with a cylindrical stem sticking out from the center of the flat side. When spun fast enough by twisting the stem between the finger and thumb, the top performs a peculiar trick: after a while it turns upside down and continues the rotation on the tail. Obviously, the potential energy of the top is rising due to such a somersault! Long ago this eccentric behavior was explained by W. Thomson[d] and since then the toy is sometimes called a Thomson's spinning top.

It turns out that an ordinary hard-boiled egg also can behave like a Thomson's top. Take an egg and twirl it as fast as possible on a flat, un-polished, hard surface, say, on smooth oilskin. After some revolutions the egg will stand up onto the end and spin about the vertical axis! Only after being noticeably slowed by friction the egg will start swaying from side to side and finally lie down onto the side. Note that for this trick you need a really *hard*-boiled egg. The soft-boiled ones won't do. The reason is that the viscous friction between layers of liquid yolk and white and between the liquid white and the shell will slow down the rotation and the egg will lose the angular momentum. This distinction in behavior offers an easy way to find out whether an egg has been boiled hard or soft. When spun, a

---

[c]Physically, egg shells at the table and metallic shells of deep-sea submarines collapse in the same manner. On the other hand, mathematically, stresses in the compressed shell are distributed similarly to tensions in the walls of a soap bubble. This analogy makes it possible to relate the critical pressure P_c to the radius of curvature R of the shell. One may simply apply the Laplace formula (see Chapter 10):

$$P_c = \frac{2\sigma_b}{R}$$

where the critical stress σ_b characterizes the strength of the shell. Thus the pressure required to break an egg is inversely proportional to the radius of the knocking end. —A. A.

[d]See page 53.

soft-boiled egg will stop after a couple of turns, whereas a hard-boiled one will keep on spinning much longer.

Now, after this extensive discourse upon boiled eggs, let us look closer at what happens to eggs during thermal treatment. What should we do, for instance, in order to prevent an egg cracking or bursting (like in a microwave oven)? When should we stop in order to cook eggs soft and not hard? Why does a knowing cook always boil eggs in salted water? The answers to these questions are not easy to find even in hefty cookbooks.

In essence, the process of egg-boiling and many other ways of heat-treating foods (such as frying, baking, etc.) are based on the fundamental process of protein denaturation. At high temperature these complex organic molecules break into fragments, changing their shape and their spatial structure. Denaturation may be caused by many factors including chemicals, enzymes, etc.

It turns out that the proteins of egg's yolk and white are different, therefore, they start denaturing at different temperatures. This detail may seem insignificant at first glance, but actually is what makes it possible to prepare soft-boiled eggs. Denaturing of the proteins in the egg white starts at $60°\,C$ ($140°$ Farenheit) whereas the yolk begins to denature only at $63°$–$65°\,C$ (or $145°$–$150°\,F$). It makes no sense to specify these temperatures precisely since proteins do not denature abruptly, but gradually. Besides, the temperatures depend on factors like the salt content of the egg, the age of the egg, etc.

Physically speaking, as soon as the egg gets into hot water a heat flow from the shell to the center arises. Problems of heat transfer in their most general form have been in the scope of mathematical physics for a long time. Hence, in order to determine how long it will take to cook a particular egg we need to know only the thermal conductivities of the egg's white and yolk and the geometrical size of the egg.

The calculation for an ellipsoidal egg was carried out by the English physicist Peter Barham. In his book *The science of cooking* he derives a formula relating the cooking time t (min) to the boiling-point of water (T_b), the egg's minor diameter d (cm), (Figure 18.2) and the initial (T_0) and intended final (T_f) temperatures of the yolk:

$$t = 0.15d^2 \log 2 \frac{(T_b - T_0)}{(T_b - T_f)}.$$

Fig. 18.2: Measuring
the egg's minor
diameter

Here log stands for the natural logarithm. For minor diameter measured in inches the overall coefficient should be substituted by 0.023.

Under the standard conditions (usually that means at normal atmospheric pressure at the sea level) the boiling-point of water is $T_b = 100° C$ ($212° F$), see Table 20.1. Therefore, according to Barham's formula, the time required to boil soft ($T_f = 63° C \sim 145° F$) a typical egg with $d = 4\ cm$, just from the fridge ($T_0 = 5° C \sim 40° F$) must be $t_1 = 3\ min\ 56\ s$. For a bigger egg with ($d = 6\ cm$) the time is twice as long: $t_2 = 8\ min\ 50\ s$[e].

Note that the formula also indicates a necessity to enhance cooking times at higher altitudes, as the boiling-point of water notably drops with altitude (see page 172). Therefore, the times recommended in books written at sea level should be appropriately extended for Alpine cooking. For instance, at an altitude of $5000\ m$ (16.5 thousand feet) water boils at $88°C$ ($190°F$). Thus, making a soft-boiled egg will take $4\ min\ 32\ s$ instead of the $3\ min\ 56\ s$ that would suffice at sea level.

Now, let us discuss salting the water for boiling eggs. The density of a fresh egg is greater than that of fresh water. So in unsalted water the egg lies on the bottom of the pan. Turbulent streams of boiling water toss the egg to and fro, knocking it against the bottom and walls, and may crack the shell[f]. The white will leak through the crack and thicken into dangling

[e]Of course, cooking is not an *exact* science and times may vary from these estimates.

[f]A greater danger comes from the expansion of the air cell in the egg. A little hole pricked in the large end will let the air out and prevent cracking. —A. A.

flakes. The meal is ruined! Yet if a prudent cook has added to the water just a half spoonful of salt, the "catastrophe" is normally avoided. Salt stimulates the white's protein denaturation. The coagulated white patches the leak like caulk.

Finally, the egg is cooked. Spoon it out of the water and try to touch while it's still wet. Certainly it is pretty hot but still possible to hold in your hands. It gets harder though, as the shell dries. Soon the egg dries completely and becomes too hot to hold. Why?

After having answered this question, try to remove the shell from the egg. You will discover that the shell has stuck in places and takes with it chunks of white. One could have prevented that by putting the freshly cooked egg into cold water. The shell and the white of the egg have different coefficients of thermal expansion; the white would have shrunk more quickly and came loose from the shell.

So far we have witnessed how the laws of mechanics, hydrodynamics, and molecular physics reveal themselves in the behavior of an ordinary egg. But could it help us to observe electric phenomena? Consider how an egg heated in a microwave oven will explode; this is due to the rapid heating of the egg by the electromagnetic field. There is another, less hazardous[g] electrical phenomenon one can observe with the help of our "protagonist". The egg shell is a dielectric. Let us follow the famous Michael Faraday who used this property in his demonstrations of static electricity.

Take a fresh egg and make two small holes in the ends: one a little bigger than another. Then prick the yolk through the hole with a long needle and stir the contents until liquid. If you now blow into the smaller hole, the liquid will flow out and leave you a hollow almost intact shell. Wash it and let dry. Now it is ready!

Try to approach it with a charged object, such as a comb which you just have used on dry hair or an ebonite stick electrified by rubbing against dry wool. Electrostatic forces will draw the empty shell and it will follow the comb like a devoted dog follows his master.

When the game is over, you may convert the empty shell into a small jet engine. Seal the bigger hole with a piece of chewing gum (everything works for a true experimentalist) and half-fill the shell with water[h]. Fix

---

[g]There are supposedly only three more words in English with their spelling ending with "dous". They all are rather common in everyday speech. Try to find them! —*D. Z.*

[h]Change the order if necessary.

this reservoir on a light carriage with freely turning wheels, as shown in Figure 18.3. Finally, put under the shell a tablet of solid alcohol and light it up. In a short while the water will come to boil and the vapor will escape through the smaller hole. According to the principles of jet propulsion this will cause our vehicle to run in the opposite direction.

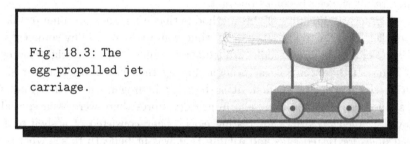

Fig. 18.3: The egg-propelled jet carriage.

We have already discussed how to recognize hard-boiled eggs from uncooked ones. But is it possible to find out without breaking whether an egg is fresh or if it has spoiled? Yes, of course. Just put it into a glass of water, Figure 18.4.

Fig. 18.4: The older the egg, the better it floats.

A fresh egg, laid a few days before, sinks right away to the bottom. An egg laid a week ago stays under water vertically but does not rise to the surface. A bad one (say, three and more weeks old) will float by the surface and even stick out a little. The thing is that as an egg "ages" its white and yolk start to degrade and dry up. As a consequence, a gas (hydrogen sulfide) is produced. Some of the gas, along with a bit of water vapor, escapes through the tiny pores in the shell, but the rest stays inside. These processes decrease the mass of the egg, while its volume remains the same. As a result the egg becomes buoyant.

A practical difficulty of this method is that it's rather tiresome to take a bucket of water to the store every time you are going to buy some eggs. A simpler way is just to look through the examined egg at a bright source of light. If the light passes through the egg then it is most likely fresh. Otherwise it must be spoiled. It is the same hydrogen sulfide that makes a spoiled egg opaque. Long ago in grocery stores there were even special devices called *ovoscopes* for testing eggs. They consisted of a shelf with oval holes for tested eggs and a lamp that was lit beneath to see whether the eggs were transparent.

In conclusion, you could try to perform this elegant trick. Put two egg-holders next to each other and place a hard-boiled egg into one of them. Now blow as hard as you can into the gap between the egg and the inner wall of the holder. The egg will jump from its holder into the next one. When after some attempts you succeed, try to explain why!

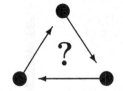

Why doesn't the gum seal on our egg-shell jet engine melt?

Chapter 19

Macaroni, spaghetti, and physics

If Narcissus was turned into a flower,
I want be metamorphosed into Macaroni.

Fellippo Sgruttendio[a].

Each and everyone knows a lot about spaghetti, and many of you have cooked it at home. But have you ever thought about the physical processes taking place in the pot where pasta is cooking, and producing a correctly boiled "pasta" (in accordance to Italian standards) as a result? Have you ever asked yourself what happens inside spaghetti when it is floating in the boiling water? Why should we always follow the cooking time instructions specified on spaghetti packaging? Why is the cooking time different for different types of pasta? How does cooking time depend on the shape of pasta (classical spaghetti with different diameters, rigatoni or bucatini, etc.)? Does the cooking time depend on the location: are you making pasta at the seaside or in the highlands? When bent, why does an uncooked spaghetti strand ("spaghetto") almost never break into two pieces but into three or four? Why do spaghetti strands never tie into knots during cooking? How do you select the type of pasta, if you already have a sauce, so that the dish will be hot and tasty?

Below we will try to gain insight into the physics of the spaghetti cooking process and find the answers to some of the questions that may arise while the interested audience waits for a pot of pasta to boil in anticipation of a hearty dinner.

---

[a]Fellippo Sgruttendio, Neapolitan poet of the eighteenth century.

19.1   A glimpse into a history of pasta and its manufacturing

Contrary to popular opinion, pasta was not brought to the West by Marco Polo after his trip to China (1295). In fact, its history began much earlier on the Mediterranean coast, at the time when the prehistoric man was leaving the nomadic lifestyle behind and started settling down and growing grain for food. The first flat cakes baked on top of hot stones were mentioned in the Old Testament (*Genesis* and *Kings*). In the first millennium B.C., Greeks were already making a thin layered pasta; they named it *"laganon"*. This word came to ancient Romans in the form of *"laganum"*, and may well have became the source for modern term *"lasagna"*.

Etruscans were also preserving thin layers of pasta. With the growth of the Roman Empire, pasta started spreading throughout Western Europe. Reliance on pasta as a way to preserve grain products emerged with the need to transfer food supplies during the periods of tribal migration. In Sicily, pasta was introduced by Arabs when they conquered the island in the tenth century. Sicilian pasta named *"trie"* may be considered the ancestor of spaghetti. It was shaped in thin strands, and the name originated from the Arabic word *"itryah"* (flat cake cut into strands). People living in Palermo started making pasta in the beginning of the second millennium. Based on a detailed will, probated by a Genoa notary public Ugolino Scarpa, we can positively argue that by 1280 macaroni products were already consumed in Liguria. It is known from the History of Italian Literature that pasta has attracted the attention of such writers as Jacopone da Todi, Cecco Angiolieri, and Felippo Sgruttendio. Ultimately, in Boccaccio's *Decameron* macaroni became a symbol of sophisticated gourmet food.

The first guilds of pasta makers (*"Pastai"*) with their own charters were created in the sixteenth century in Italy, where they received political and public recognition. At that time macaroni was regarded as a food for the rich, especially in the provinces which did not grow their own durum wheat (e.g., Naples). The invention of the mechanical press resulted in lowered production costs and hence the price of the product. As a result, by the seventeenth century pasta turned into one of staple foods consumed by all social classes; now it is widespread in all countries of the Mediterranean basin. Naples has become a major center of pasta manufacturing and export. Here, pasta with basil tomato sauce or sprinkled with grated cheese is sold on every street corner. In Northern Italy pasta became popular at the end of the eighteenth century mainly due to Pietro Barilla, who opened a small factory in Parma and later became a major producer in the Italian food industry.

Modern methods of pasta production are primarily based on the process of extrusion (pushing out through holes) and dragging. Extrusion was invented and used for the first time in the manufacture of long metal components with specified cross sections (see Figure 19.1). Extrusion processes are based on the property of fluidity and on subsequent pushing of material through a rigid die by means of compression. It can work both in cold and hot conditions. Dragging is a process similar to extrusion, the only difference being that in the case of dragging the material is pressed through the die located at the vessel's outlet and thus it becomes a process of stretching rather than compression. This method is used by the metal-processing industry for cylinder, wire, and the pipe manufacturing. It allows reduction of the diameter of metal wire to 0.025 *mm*. Other materials that can be processed by extrusion are: polymeric compounds, ceramics, and food. Dies used to produce spaghetti are shown in Figure 19.2.

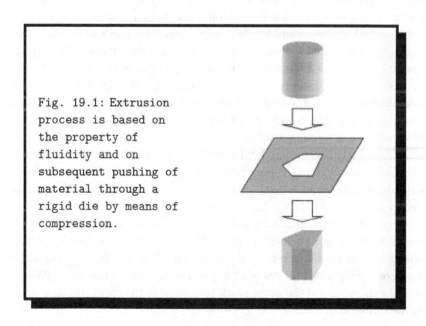

Fig. 19.1: Extrusion process is based on the property of fluidity and on subsequent pushing of material through a rigid die by means of compression.

Fig. 19.2: Spaghetti
die.

19.2 A scientific way of spaghetti cooking

Before we begin, let us try to understand what is involved in the process
of pasta cooking. In the flour, molecules of starch are grouped into granules with diameter of 10–30 microns, which, in turn, are surrounded with
different proteins. In the process of pasta fabrication two of them, gliadin
and glutenin, combine with water, unite, and form a continuous net, called
gluten, which is strong and has low permeability for water molecules. This
net covers the starch granules. Cooking time is directly related to the capacity of the starch molecules (surrounded by gluten in the process of pasta
drying) to absorb water, which begins penetrating through the gluten net-
work and diffuses to the center of spaghetto as soon as the pasta is placed
in the pan of boiling water. At a temperature of about $T_g = 70°C$, starch
molecules begin forming a gel-like compound, which hinders water absorption. Spaghetti is considered "al dente" (to the tooth) the moment when
the gel-like starch in pasta absorbs the minimum amount of water necessary
to make it sufficiently soft. Hence in order to cook pasta it is necessary to
deliver hot water inside the initially dry spaghetto.

As a result of the above discussion, the physicist can formulate a simple
model of the pasta cooking process. Consider an homogeneous cylinder
(spaghetto) of the diameter d and diffusion coefficient $\mathcal{D}$ that is placed into
a pan of boiling water (at sea level it happens at temperature $T_b = 100°C$);
at what time, τ_0, will the water reach the center of the spaghetto as a result
of the diffusion process[b]?

---

[b]Here we ignore the decrease of water temperature in initially cold spaghetto. The
process of heat transfer is faster then diffusion and in any case is described by the

The process of diffusion is described by the complex differential equation in partial derivatives:

$$\frac{\partial n\left(\mathbf{r},t\right)}{\partial t} = -\mathcal{D}\frac{\partial^2 n\left(\mathbf{r},t\right)}{\partial \mathbf{r}^2}, \tag{19.1}$$

where $n\left(\mathbf{r},t\right)$ is the value of water concentration at the moment t in the point $\mathbf{r} = \left(x,y,z\right)$. The fact that the external temperature is fixed as T_{b} means that the surface temperature of the cylinder is equal to T_{b} at any moment, while the time τ_0 corresponds to the moment when the water concentration at the axis of the cylinder reaches the value n_0.

In principle, equation (19.1) can be solved exactly. Nevertheless, we assume that our reader may not be familiar with the theory of differential equations and we will show how to get the necessary formula for τ_0 using the method of dimensional analysis. Let us look at the equation (19.1). The comparison of dimensionality of the left- and right-hand sides gives

$$\frac{[n]}{[t]} = [\mathcal{D}]\frac{[n]}{[r]^2}, \tag{19.2}$$

or

$$[t] = [\mathcal{D}]^{-1}[r]^2. \tag{19.3}$$

The only physical value with the dimensionality of length in the problem under discussion is that of the cylinder diameter d. The dimensionality of the diffusion coefficient $\mathcal{D}$ is m^2/s and we do not have any other sources of time dimensionality on the right-hand side of equation (19.3). Hence, the diffusion coefficient in the requested expression for τ_0 should enter as $\mathcal{D}^{-1}$. As the consequence, the diameter d appears in the expression for τ_0 as d^2 :

$$\tau_0 = ad^2. \tag{19.4}$$

The coefficient a has the dimensionality of the inverse diffusion coefficient. It is mainly determined by the value of $\mathcal{D}$, but also depends on the dimensionless ratio of temperatures $T_{\mathrm{b}}/T_{\mathrm{g}}{}^{\mathrm{c}}$. The starch gelification temperature, T_{g}, is constant while the boiling temperature, T_{b}, depends on

equation of the same type.

[c] As was mentioned in the previous footnote, we ignore the process of heat conductivity of the cylinder and consider only the diffusion of hot water. Hence the initial temperature of spaghetto, which is at room temperature, is irrelevant.

the height with respect to the sea level, H. Consequently, the coefficient $a = a(H)$ depends on H and hence, cooking time $\tau_{sp} = \tau_{sp}(H)$. The recommended cooking time on the packaging corresponds to the sea level, where $T_b = 100°C$. At high altitudes, where water boils at lower temperatures, cooking time should be extended. In the extreme case of Everest (its height is 8,848 m) $T_b = 73°C$, which is very close to T_g and pasta can not be cooked at all.

We have found the time it takes for water to reach the central part of the spaghetto. Now, in order to take into account the nationality of the pasta eater and to leave him the freedom to have the spaghetti either "al dente" or "well done", we add to the final formula for cooking time τ_{sp} some constant b:

$$\tau_{sp} = ad^2 + b. \tag{19.5}$$

Indeed, when the first term in equation (19.5) determines the time of hot water diffusion to the core of the spaghetto, the second one tells us to what extent the starch molecules there have formed a gel-like compound. That is why for Italians, who prefer to eat spaghetti "al dente", the process of the starch gelification will not be finished through the entire volume of a whole spaghetto. The core of the spaghetto remains relatively firm. As a result, the coefficient b turns out to be negative for them. Spaghetti eaters in some other nations believe that pasta has to be well done and their cooking time noticeably exceeds the time written on the Italian spaghetti packaging[d].

Let us now go to a supermarket and buy every kind of cylindrically shaped pasta: capellini, spaghettini, spaghetti, vermicelli, bucatini. Read and collect in Table 19.2 the recommended cooking times (the column

---

[d]Let us mention that the obtained result for spaghetti cooking time τ_{sp} presents itself as a special case of the well-known Fick law for dependence of cooking time for a spherically symmetric object (piece of meat, Thanksgiving turkey) on its mass M

$$t_c \sim M^{2/3}.$$

Indeed, the formula (19.1) was studied by means of dimensional analysis of the diffusion equation. The latter has the same form as the heat transfer equation (with substitution of the diffusion coefficient by the heat conductivity and local concentration by local temperature). Such analysis could also be applied successfully to the case of a spherically symmetric object and to give the same result as the Fick law for cooking time:

$$\tau_c \sim d^2 \sim M^{2/3}.$$

"Cooking time experimental". Then take a slide gauge, measure the corresponding diameters and fill out the same table (the column "diameter, external/internal"):

Table 19.1

Type of pasta	Diameter, external/internal (mm)	Cooking time experimental (min)
capellini n.1	1.15/-	3
spaghettini n.3	1.45/-	5
spaghetti n.5	1.75/-	8
vermicelli n.7	1.90/-	11
vermicelli n.8	2.10/-	13
bucatini	2.70/1	8

In order to find the numerical values of the coefficients a and b it is enough to write the equation (19.5) using the data from two rows of our Table 19.2:

$$t_1 = ad_1^2 + b$$
$$t_2 = ad_2^2 + b \qquad (19.6)$$

and to solve the system of these two simple equations. As the reference case, we choose the data for spaghettini n.3 and vermicelli n.8. This gives:

$$a = \frac{t_2 - t_1}{d_2^2 - d_1^2} = 3.4\,\mathrm{min}\,/mm^2$$

$$b = \frac{d_2^2 t_1 - d_1^2 t_2}{d_2^2 - d_1^2} = -2.3\,\mathrm{min}\,.$$

We bought the Italian pasta, and the recommended cooking times on the packaging are given to obtain the "pasta al dente". Consequently, one can see that the coefficient b for Italians is indeed negative.

Having the numerical values of the coefficients a and b we can check how our formula works for other types of cylinder-shaped pasta. The results of our calculations can be found in Table 19.2 and show very good agreement with the experimental data for all rows except the two extremes: capellini and bucatini.

Table 19.2

Type of pasta	Cooking time experimental (min)	Cooking time theoretical (min)
capellini n.1	3	2.2
spaghettini n.3	5	5.0
spaghetti n.5	8	8.1
vermicelli n.7	11	10.0
vermicelli n.8	13	13.0
bucatini	8	22.5 ?!

This is a typical situation in theoretical physics: predictions of a theory have a range of validity corresponding to the assumed simplifications made for this theory. For instance, let us take a look at the last row of the table: for bucatini the difference between the theoretical and experimental values is striking: 22.5 min versus 8 min. This contradiction reflects an important fact: the range of possible thicknesses for all varieties of a whole cylinder-shaped pasta (capellini, spaghetti, vermicelli) is very narrow: 1 *mm* only. Indeed, our calculation of the "al dente" cooking time for bucatini in the approximation of the uniform cylinder gave 22.5 min, which would completely boil the periphery of such a thick "spaghettone" into mush.

The way to make such a thick spaghetti edible was found empirically: a hole should be made in the spaghetti strand along its axis. In the process of cooking, water enters through this hole, and there is no longer a need to deliver it to the core from the outside. One can try to modify the formula (19.5) deducting the internal diameter from the external one, and immediately the theoretical result comes close to the reality:

$$t = a\left(d_{\text{ext}} - d_{\text{int}}\right)^2 + b \approx 7.5 \min.$$

Nevertheless, one should keep in mind that the hole in the pasta can not be less than ~ 1 mm in diameter, otherwise, due to capillary pressure,

$$P_{\text{cap}} = \frac{4\sigma\left(T = 100°C\right)}{d_{\text{int}}} \sim 200 Pa,$$

water will not be able enter into it. The value $200Pa$ corresponds to the pressure a couple of centimeters of water above the cooking pasta.

Another deviation of theory from reality occurs for very thin pasta. The reason for this error is obvious: speaking about cooking pasta "al dente" we have chosen the parameter b as negative: $b = -2.3 \, \text{min}$. Formally, this means there should exist a pasta thin enough to not even need cooking at all to be eaten "al dente". The corresponding critical diameter d_{cr} can be determined from the relation:

$$\tau_{\text{cr}} = ad_{\text{cr}}^2 + b = 0,$$

which gives

$$d_{\text{cr}} = \sqrt{|b|/a} \approx 0.82mm.$$

One can see that the real diameter of capellini (1.15 mm) is not far from this critical value, so the underestimated capellini cooking time in Table 19.2 is the result of this limitation of our model.

19.3 Spaghetti knotting

Cooked spaghetti strands entangled with each other present a complex clew in hot water, however, the authors of this book have never seen them knot themselves. The reasons why this does not happen can be learned from a new field in statistical mechanics: the statistics of polymers.

The probability of a long polymer chain self-knotting is determined by the expression[e]

$$w = 1 \quad \text{опр} \left(\frac{L}{\gamma \xi} \right),$$

where L is the full length of a polymer, ξ is the characteristic length at which the polymer can change its direction by $\pi/2$ and $\gamma \approx 300$ is the large factor, obtained as a result of numeric modeling and in some theoretical models. Applying this formula to spaghetti, where $\xi \approx 3 \, cm$, one can find the length $L_{\min}$ when the probability of self-knotting becomes noticeable ($w \sim 0.1$) :

$$\exp \left(-\frac{L_{\min}}{\gamma \xi} \right) \approx 0.9$$

---

[e]A.V. is grateful to A.Y. Grosberg for introduction to the theory of knots.

which gives

$$L_{\min} \approx \gamma \xi \ln 1.1 \approx 30\xi \approx 1 \ m.$$

The length of a standard spaghetto is 23 *cm* and this is not long enough to form knots.

19.4 Breaking a spaghetti strand

In the beginning of this chapter we mentioned another interesting property of spaghetti related to its mechanical fracture. Take both ends of a spaghetto and bend it into an arch, gradually increasing its curvature. One can suppose that sooner or later it will break into two parts somewhere near the middle. It turns out, though, that in this case our intuition is not correct: nearly always it will break into three or more pieces.

Such unusual behavior of a spaghetti strand attracted the attention of numerous scientists, Richard Feynman among them. But only a short time ago in 2005, owing to the research studies conducted by the two French physicists Auduly and Neukirch, a quantitative description of this phenomenon was obtained.

The scientists studied the behavior of a thin, elastic rod under the effect of flexural deformation. They wrote the differential equation which describes the distribution of the tension (so-called Kirchhoff equation) in a curved elastic rod, first with both ends fixed. Then they studied what will happen to the tension distribution along the rod after the instantaneous release of one of its ends. Only a numeric solution was obtained, but it provided an understanding of the essence of the process. Qualitatively, the explanation is as follows.

Let us suppose that due to the applied mechanical stress, the first fracture occurs at some (weakest) point of the strand. It would seem that after being broken, both parts of the rod should return to their equilibrium positions. This is true, however the transition to the equilibrium state occurs in a very nontrivial way. The first fracture generates flexural waves in both fragments of the rod which start to propagate along each of the fragments. Evidently, flexural waves of this kind (generated by the first fracture) will dampen out with time, but, at certain ratios between the rod length and its elastic modulus, the wave propagation can lead to a subsequent rod fragmentation. Indeed, propagation of such a wave means periodic growth and decrease of the local flexural stress along the rod. It is important to note that these flexural waves propagate on the background of the already existing initial homogeneous bend of the rods, which relaxes much more slowly than the flexural wave period. As a result of the summation of these two quasistatic and dynamic stresses, further rod fractures may occur at other points, where such sums exceed a critical value. It is also noteworthy that after intricate computations, the researchers confirmed their theoretical conclusions by filming the experimental spaghetti break studies with a high-speed camcorder (Figure 19.3).

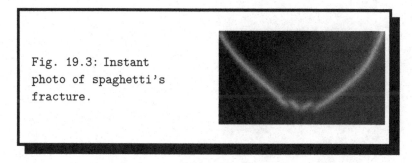

Fig. 19.3: Instant photo of spaghetti's fracture.

While the spaghetti is cooking, you can entertain yourself with several remaining dry spaghetti by experimentally verifying the findings of Audoly and Neukirch.

Chapter 20

Waiting for the kettle to boil

A bright idea came into Alice's head.
"Is that the reason so many tea-things are put out here?" — she asked.
"Yes, that's it," — said the Hatter with a sigh: "it's always tea-time, and we've no time to wash the things between whiles."

Lewis Carroll, *"Alice's Adventures in Wonderland."*

There are thick Eastern manuscripts as well as long, detailed chapters in special books devoted entirely to the tea drinking ritual. Yet when taking another, unconventional, peek at the process, one surely finds interesting and edifying physical phenomena galore, which are not described even in the most reverent culinary "oracles".

To limber up a little, let's perform the following experiment. We will take two identical tea-kettles, each with an equal amount of cold water (same initial temperature in both), and put them on the burners with the same heating power. One of the subject kettles will be covered with the lid, and the other one will remain bareheaded. Which will boil first? Any housewife (no offence to the intelligence of housewives is intended) will give you the correct answer right away. If she wants to have hot water faster, she will put the lid on and reply that the water will boil first in the covered kettle. Well, not taking this statement for granted, let's check it as we are supposed to, experimentally, and wait until the water does indeed start to boil, then have a discussion on the resulting observation afterwards.

In the meantime, while our two kettles are getting hot, we put one more, identical kettle on a third burner. The volume of water and its initial temperature again are the same as for the two previous kettles, as well as the power of the burner. Now, we aim to get the water in this last tea-kettle boiling somehow faster than in the other two. How could we possibly raise the water temperature in this kettle faster? A trivial way would be to stick an extra heating coil in it. Let's say we do not have any available. Perhaps we need to just add some hotter water to the kettle to reach the boiling point faster? It turns out that on the contrary, it will slow it down. To prove this, let us assume that the original amount of water of mass m_1 at a temperature T_1 did not mix with the added water (m_2 at T_2) and did not exchange any heat with this new mass of water either. The amount of heat required for the original volume of water in the kettle to boil would be $Q_1 = c\,m_1\,(T_b - T_1)$ where c is the specific heat of water. But now, besides this energy, the mass m_2 has to be heated to the same boiling temperature, so that the total amount of heat needed is:

$$Q_1 = c\,m_1\,(T_b - T_1) + c\,m_2\,(T_b - T_2).$$

Even if we pour boiling water straight into the tea-kettle, the added water will cool down somewhat during the filling due to heat transfer, lowering its temperature below T_b. Obviously our quite naive conjecture that the two portions of water in the same pot remain unmixed after the addition did not affect in any way the law of conservation of energy in the system, yet permitted us to handle the evaluation faster and more simply.

By the time we've finally given up the idea of adding water into the third kettle, the first of our two "original" kettles begins hissing. And what is the physical mechanism behind this familiar sibilant sound, and what is its characteristic frequency? Next, we shall try to answer these questions.

As the first candidate to generate this whistling discord, one could suggest oscillations excited in the liquid when the steam bubbles take off from the walls and bottom of the reservoir (the kettle in this case). These bubbles always start developing on all sorts of microcracks and other defects, ever present on any real surface. The typical size of such bubbles, before the water starts to boil, is about $1\,mm$ (after that they can reach as much as $1\,cm$). To estimate the frequency of sound produced by the simmering liquid, we need to know how long it takes for the bubbles to release from the bottom. This time actually measures the length of the push that the liquid experiences when each bubble takes off, and therefore, the period of the vibrations they excite. Within our assumptions, the sought-after frequency is determined by the reciprocal of this time: $\nu \sim \tau^{-1}$.

When the nascent bubbles are resting on the bottom, there are two forces acting upon them[a]: the Archimedean buoyancy force, $F_A = \rho_w \, g \, V_b$ (here V_b is the volume of the bubble and ρ_w is the density of water) pushes it upward, and the surface tension that keeps it attached to the surface, $F_S = \sigma \, l$ (where l is the length of the contact area between the bubble and the surface). As the bubble grows (V_b increases), the Archimedean force is rising too, and at a certain moment, it exceeds the retaining force of surface tension. The bubble takes off, starting on its journey upward, Figure 20.1. Hence, the resulting force acting on the bubble during its "departure" stage should be of the order of F_A. Whereas, the acceleration of a bubble in liquid is defined, of course, not by its own negligible mass (consisting mostly of the mass of air trapped in it), but by the mass of liquid involved in the motion. For a spherical bubble this so-called *associated mass* equals $m^* = \frac{2}{3} \pi \, \rho_w \, r_0^3 = \frac{1}{2} \rho_w \, V_b$ (r_0 being the bubble's radius).

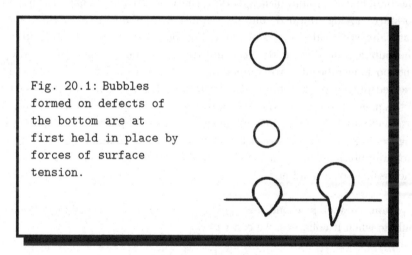

Fig. 20.1: Bubbles
formed on defects of
the bottom are at
first held in place by
forces of surface
tension.

Thus, for the acceleration of the bubble during the initial stage, one finds:

$$a \sim \frac{F_A}{m^*} = 2g.$$

Now we can evaluate the bubble release time, considering (again, for

[a]Here we disregard the tiny weight of the bubble.

simplicity's sake) the motion to be uniformly accelerated. Our subject bubble will climb up to the height comparable to, say, its size within the time of

$$\tau_1 \sim \sqrt{\frac{2r_0}{a}} \sim 10^{-2}\, s. \tag{20.1}$$

Then, the corresponding characteristic frequency of the sound generated at the bubble's take off should be equal to $\nu_1 \sim \tau_1^{-1} \sim 100\, Hz$. This seems like maybe an order of magnitude less[b] than the sibilant tone one hears when a tea-kettle is being heated on a stove (long before the water starts actually boiling)[c].

So it turns out that there must be another cause for the tea-kettle's hissing when it's warming up. To establish this second reason, one would have to closely follow the bubble's fate after it leaves its parental surface. Having taken off from the hot wall (or bottom) where the vapor pressure in the bubble was around atmospheric (otherwise it couldn't expand enough to start moving upward), our protagonist hurries up in the higher, and, naturally, still colder, layers of water. So, the saturated water vapor, filling the bubble, cools down too, causing the inside bubble pressure to drop and no longer be able to compensate for the external pressure of liquid exerted on the poor bubble. As the result, the squashed bubble flops or gets squeezed into a tiny one (the latter happens if, besides water vapor, it also had a little bit of air inside), generating a sound pulse in the liquid (see Figure 20.2). This process of the massive death or maiming of numerous steam bubbles on their way to the water surface is indeed what we hear as the hissing noise. And now, we will try to estimate its frequency, of course.

From Newton's second law, for the mass m of water rushing into the bubble when it collapses, we can write:

$$m\, a_r = F_p = S\, \Delta P.$$

Here $S = 4\pi r^2$ is the surface area of the bubble, F_p is the total pressure

---

[b]Frequencies in the $100\, Hz$ range are familiar from the humming noise of old radios.— A. A.

[c]Note that the surface tension did not make it into (20.1) indicating, in a way, that the bubbles produce sound not only when leaving the surface but also during their accelerated motion upwards. This lasts until the buoyancy force gets compensated by the viscous friction, which is proportional to velocity.

Fig. 20.2: Before the active boiling starts hundreds of tiny bubbles collapse making the tea-kettle sing.

squeezing the bubble, ΔP is the pressure difference across the bubble's envelope, and a_r is the inward acceleration of the envelope. It's pretty clear that the mass involved in the process of "squeezing" should be of the same order as the product of water density times the bubble's volume: $m \sim \rho_w r^3$. So, we could rearrange Newton's equation in the following manner:

$$\rho_w \, r^3 \, a_r \sim r^2 \, \Delta P.$$

Further, neglecting the fraction of pressure arising from the bubble's surface curvature, as well as the smidgen of air possibly ensnared inside the bubble, we will consider ΔP to be constant (more precisely, depending only on the temperature difference between the bottom and surface layers of water in our tea-kettle). Now, evaluating the acceleration $a_r = r'' \sim r_0 / \tau_2^2$ where τ_2 is the "flop" time we are looking for, we find that

$$\rho_w \frac{r_0^2}{\tau_2^2} \sim \Delta P,$$

which gives

$$\tau_2 \sim r_0 \sqrt{\frac{\rho_w}{\Delta P}}. \tag{20.2}$$

Near $T_b = 100° \, C$, the saturated water vapor pressure drops by about $3 \cdot 10^3 \, Pa$ per degree Celsius of the falling temperature (see Table 20.1). Hence, we could assume $\Delta P \sim 10^3 \, Pa$, and finally write for time $\tau_2 \sim 10^{-3} \, s$, leading to the noise frequency of $\nu_2 \sim \tau_2^{-1} \sim 10^3 \, Hz$. This answer is already much closer to the value perceived by our ears[d].

Table 20.1: TEMPERATURE DEPENDENCE OF PRESSURE OF SATURATED
WATER VAPOR.

Temperature, $°C$	96.18	99.1	99.6	99.9	100	101	110.8
Pressure, kPa	88.26	98.07	100	101	101.3	105	147

One more fact supporting our conclusion that the proposed mechanism is indeed responsible for the tea-kettle noise is that, according to equation (20.2), its characteristic (high) frequency drops as the temperature grows. Right before the boiling begins, the bubbles cease flopping even in the upper layers of water. Then the only remaining sound becomes that produced by the bubbles taking off from the bottom. The frequency of the "tune" drops noticeably when the water in the kettle is about to start boiling. After it finally happens though, the kettle's "voice" may change again, especially if one opens the lid: the gurgling sound we hear is generated now by the bubbles rupturing right on the water's surface. This pitch depends also on the water level, as well as on the kettle's shape[e].

Thus, we have established that the kettle's noise before the water starts boiling is related to the hundreds of steam bubbles produced on the hot bottom departing to the surface, and then perishing in the cooler upper layers of water. All these processes become especially apparent if one is heating water in a glass pot with transparent walls. Let's not hurry though to congratulate ourselves that we were the first to sort things out on this interesting question of singing boiling water. Back in the eighteenth century, the Scottish physicist Joseph Black[f] was studying the phenomenon and had established that the sound was produced by a duet consisting of the steam bubbles ascending to the surface and the vibration of the vessel's walls.

[d] According to Figure 20.3 the temperature drop (and the corresponding pressure difference ΔP and frequency ν) may be up to the order of magnitude higher. —A. A.

[e] One more argument in favor of the proposed mechanism is that bubbles in fizzy drinks don't make the sound we have been talking about. The difference with boiling water is that the carbon-dioxide filled bubbles can not collapse. —A. A.

[f] J. Black, (1728–1799), Scottish physicist and chemist, he was the first to point out the difference between heat and temperature. He also introduced the idea of heat capacity.

By now the water in the first of our subject tea-kettles (the one covered with a lid, remember?), the predicted favorite winner, has started to boil. The moment is quite resolutely announced by the stream of steam jetting out of the kettle's spout. And what could be the velocity of the stream, by the way? One can solve this (honestly, not the most challenging) problem by noticing that during the steady boil, all the energy provided to the kettle is being spent on vaporizing the water. Let's assume that in this case the only way steam is escaping is through the spout. Suppose further that a mass ΔM of water is vaporized in time Δt, at the expense of the supplied heat. Then, one could write the following balance equation:

$$r \, \Delta M = \mathcal{P} \, \Delta t,$$

where r is the specific heat of vaporization (heat per unit mass), and $\mathcal{P}$ is the power of the heater. During the same time Δt, the same mass ΔM is supposed to leave the kettle through its spout since otherwise the vapor would accumulate under the lid. If the area of the outlet (pretty much the perpendicular cross section of the spout) equals s, the steam density is $\rho_s(T_b)$, and v is the velocity in question, then the following relation holds:

$$\Delta M = \rho_s(T_b) \, s \, v \, \Delta t.$$

From the previous table, we pick $\rho_s(T_b) = 0.6 \, kg/m^3$ as the density of saturated water vapor at $T_b = 373° \, C$. If one doesn't happen to have a suitable table handy, one can use the Clapeyron–Mendeleev gas law[g]:

$$\rho_s(T_b) = \frac{P_s(T_b) \, \mu_{H_2 O}}{R \, T_b} \approx 0.6 \, kg/m^3. \tag{20.3}$$

Thus, for the velocity of steam escaping the spout, one finds:

$$v = \frac{\mathcal{P} \, R \, T_b}{r \, P_s(T_b) \, \mu_{H_2 O} \, s}.$$

[g]The expression for gas density ρ can be obtained from the equation of state of the ideal gas $PV = \frac{m}{\mu} RT$, where P, T, and m are the pressure, temperature, and mass of gas enclosed in the volume V respectively. This can be easily brought to the form

$$\rho = \frac{m}{V} = \frac{P \mu}{RT},$$

where μ is the molecular mass of the gas and R is the gas constant per mole.

After the required substitutions: $\mathcal{P} = 500\,W$, $s = 2\,cm^2$, $r = 2.26\cdot10^5\,J/Kg$, $P_s(T_b) = 10^6\,Pa$, and $R = 8.31\,J/K \cdot mole$, we find the velocity to be $v \sim 1\,m/s$.

Now, finally, the water is boiling in the second (open) tea-kettle. It has noticeably lagged behind the capped victor. One should be very careful taking this one off the stove; if you just grab it by the handle, you can very easily scald yourself (we are sure, though, that our reader understands that safety is first in any kind of experiment). Anyway, our next question is a safety-related one: which would scald worse, the steam or the boiling water? If we add some required physical constraints to this question, we could paraphrase it in the following way: which would scald worse, a certain mass of steam or the identical mass of boiling water?

Suppose there was $V_1 = 1\,l$ of saturated, $100°\,C$, steam under the lid of the tea-kettle. Imagine further that after the lid is opened, one tenth of this steam condenses on an unlucky hand. We know already that the density of water vapor at $T_b = 100°\,C$ equals $0.6\,Kg/m^3$. Hence, the mass of water condensed on the hand is going to be $m_s \approx 0.06\,g$. The amount of heat produced during the condensation and subsequent cooling of the condensed water from $100°\,C$ down to the room temperature $T_0 \approx 20°\,C$ then will be $\Delta Q = r\,m_s + c\,m_s(T_b - T_0)$. And, therefore, as $c = 4.19\,J/Kg$ and $r \approx c \cdot 540°\,C$, it takes about ten times more of the boiling water than the hot steam to get the same thermal effect! Besides, the area affected by the steam (due to the much higher mobility of the vapor molecules) is always significantly larger than that of hot water poured on a surface. Thus, the answer to our question is that the steam would be unequivocally much more dangerous than water of the same (boiling) temperature.

However, performing all these dangerous evaluations, we have swerved quite a bit from our original test with the two tea-kettles. Why did it take water in the open one so much longer to start boiling in the first place? Let's look at the phenomenon more closely. The answer seems to be almost trivial: during the heating, the nimblest molecules of water (those with higher velocity) can easily escape from the open kettle, pilfering some energy from the water remaining inside and, by doing so, effectively cooling it (this process is nothing but evaporation). Hence the heater in this case is supposed to provide not only energy to heat the water to boiling temperature, but also the heat required to evaporate some of the water. Thus, it's clear that it is going to take more energy (and, therefore more time because the heating power is fixed) than just heating the water in the capped kettle, where those swift fugitives have no other options but to congregate under the lid, building up the saturated water vapor, and eventually return to water, bringing along the stolen energy surplus.

Yet, there are two more effects occurring at the same time, although opposite to the one above. First, during the evaporation, the mass of water which needs to be warmed to T_b drops somewhat; secondly, in the open vessel, the pressure on the water is atmospheric and, hence, the boiling process starts at exactly $100°\,C$. On the other hand, in the covered kettle, if it's filled so that the steam can't make it out through the spout, the pressure on the water surface will rise because of the intensive vaporizing. Note that now it's the sum of the partial pressures of a small amount of air under the lid as well as of the steam itself. With increasing external pressure, the boiling temperature should rise too, for it is determined by the equality of the saturated vapor pressure inside the bubbles to the external pressure. So, which of these factors should we prefer as the decisive one?

Each time such an uncertainty comes, one should retreat to a precise calculation or, at least, evaluate the magnitudes of the effects involved. So, first, let's estimate the amount of water which escapes from the open kettle before it starts boiling.

Molecules in liquids strongly interact with each other. In crystals, the potential energy of molecules is much greater than their kinetic energy, in gases, the kinetic part dominates. In liquid, the potential and kinetic energies are of the same order. So, molecules in the liquid state most of the time fluctuate around some "ascribed" equilibrium positions, yet once in a while manage to hop to a different neighboring equilibrium location. "Once in a while" just means a far longer duration compared to the period of oscillations about equilibrium points. However, in our conventional time scale, such hops occur quite frequently: in just one second, a jittery molecule of liquid can change its equilibrium position billions of times!

However, not every single molecule which happens to be wandering by the surface of the liquid can actually escape from it. To finally free themselves, such molecules ought to spend some energy performing work against the interaction forces. One may say that the potential energy of a molecule of water is less than that of a molecule of steam by the amount of the heat of evaporation, normalized per single molecule. Then, if r is the specific heat of evaporation, the molar evaporation heat is μr, and the "molecular" heat of evaporation will be $U_0 = \mu r / N_A$ (N_A being Avogadro's number). This work is performed at the expense of the kinetic energy of the molecule's thermal motion E_k. The corresponding average value $\bar{E}_k \approx kT$ ($k = 1.38 \cdot 10^{-23} J / K$ is the Boltzmann[h] constant) turns out to be far less than U_0. Nevertheless, according to the laws of molecular physics there will always be some number of molecules with kinetic energies high enough to overcome the attraction forces and flee away. The concentration of these extremely swift molecules is given by the following expression:

$$n_{E_k > U_0} = n_0 \, e^{-\frac{U_0}{kT}}, \qquad (20.4)$$

where n_0 is the total density of molecules, and $e = 2.7182\ldots$ is the base of natural logarithms.

Now for the time being, we shall forget about the hops of molecules in liquid, and consider these high-energy molecules as a gas. A molecule of such a gas can reach the surface from inside in a short instant Δt provided its velocity v is directed outwards and it has started less than $v \, \Delta t$ away from the surface. For a surface area S, these are the molecules from the cylinder of the height $v \, \Delta t$ with the base S. Let us assume for simplicity that $\sim \frac{1}{6}$ of all the molecules in the cylinder (that is $\Delta N \sim \frac{1}{6} n S v \, \Delta t$) move towards the surface. Taking the density of the molecules with energies greater than U_0 from equation (20.4) we obtain for the evaporation rate (that is the number of molecules that escape the liquid in unit time):

$$\frac{\Delta N}{\Delta t} \sim \frac{n \, S \, v \, \Delta t}{6 \, \Delta t} \sim S \, n_0 \sqrt{\frac{U_0}{m_0}} \, e^{-\frac{U_0}{kT}},$$

where we have taken the velocity $v \sim \sqrt{U_0 / m_0}$. So, the mass carried away from the liquid per unit time is equal to:

$$\frac{\Delta m}{\Delta t} \sim m_0 \, \frac{\Delta N}{\Delta t} \sim m_0 \, S \, n_0 \sqrt{\frac{U_0}{m_0}} \, e^{-\frac{U_0}{kT}}. \qquad (20.5)$$

It becomes more useful to recalculate this mass normalizing it to the $1 \, K$ temperature increase while the kettle is being heated. To do so, we will use the energy conservation law: our kettle, in a time Δt, receives the amount of heat from the burner $\Delta Q = \mathcal{P} \, \Delta t$ ($\mathcal{P}$ is the burner's power),

[h]L. Boltzmann, (1844–1906), Austrian physicist; one of the founders of classical statistical physics.

and, consequently, the water temperature rises by ΔT, entailing:

$$\mathcal{P} \Delta t = c M \Delta T,$$

where M is the mass of water in the kettle (here we disregard the heat capacity of the kettle itself). After plugging $\Delta t = c M \Delta T / \mathcal{P}$ into the equation for evaporation rate (20.5) we have:

$$\frac{\Delta m}{\Delta T} = \frac{\rho c S M}{\mathcal{P}} \sqrt{\frac{r \mu_{H_2 O}}{N_A m_0}} \, e^{-\frac{r \mu_{H_2 O}}{N_A k T}} = \frac{\rho c S M \sqrt{r}}{\mathcal{P}} e^{-\frac{r \mu_{H_2 O}}{N_A k T}}.$$

As the tea-kettle is being heated, the temperature increases from room temperature to the boiling point, $373\,K$. However, we conjecture at this point (and rightfully) that the majority of the mass is being lost while the temperature of water is already pretty high (close to its boiling value) so that we could put, say, $\bar{T} = 350\,K$ as the average temperature into our exponential expression above. For the rest of the terms we assume: $\Delta T = T_b - T_0 = 80\,K$, $S \sim 10^{-3}\,m^2$, $\rho \sim 10^3\,Kg/m^3$, $\mu_{H_2 O} = 0.018\,Kg/mol$, and $c = 4.19 \cdot 10^3\,J/Kg$. After putting all these values into the formula, we finally find:

$$\frac{\Delta m}{M} \approx \frac{\rho c S}{\mathcal{P}} \sqrt{r} \, e^{-\frac{r \mu_{H_2 O}}{R \bar{T}}} (T_b - T_0) \approx 3\,\%.$$

Thus, while being warmed up to the boiling temperature, just a small percentage of the total mass of water actually leaves the tea-kettle. The evaporation of such a mass takes extra energy from the heater, and naturally protracts the heating before it reaches the boiling point. To understand by how much, one could calculate and find out that the vaporization of this volume of water requires an amount of energy from the heater equivalent to heating from room temperature to boiling point about one fourth of the total mass of water in the tea-kettle.

Now let's turn to the second (or maybe the first – we don't remember already) covered kettle, and look more closely at the effects inhibiting the boiling. The first of them (potential change of mass of water during the heating) should have dropped off right away for, as we have just shown, the evaporation of about 3 percent of the water is energetically equivalent to the heating of about 25 percent of the water and, therefore, the heat required to bring this extra 3 percent of the water mass to the boil in the closed tea-kettle can be disregarded.

It turns out that the second of the effects (the increase of pressure over the water in the covered vessel) can not play any noticeable competitive role either. Indeed, if the tea-kettle is completely filled with water (steam can't escape from the spout), then the additional (to the atmospheric) pressure obviously can not exceed the lid's weight divided by its area, for otherwise the lid would start to bounce, releasing the steam. Assuming $m_{lid} = 0.3\,Kg$ and $S_{lid} \sim 10^2\,cm^2$, we can limit this extra pressure by:

$$\Delta P \leq \frac{m_{lid}\,g}{S_{lid}} \approx \frac{3\,N}{10^{-2}\,m^2} = 3 \cdot 10^2\,Pa.$$

Having checked once more in Table 20.1, one finds that such an increase shifts the boiling temperature by no more than just $\delta T_b \sim 0.5°\,C$. Hence, to make the water boil it would take an extra heating energy of $\delta Q = c\,M\,\delta T_b$. Comparing this to $r\,\Delta M$, we see that the inequality $r\,\Delta M \gg c\,M\,\delta T_b$ holds with the safe ratio of at least as much as 30:1 allowing us to conclude that the increase of the boiling point of the water in the fully filled tea-kettle covered with the lid can not seriously compete (regarded energywise) with the evaporation of water from the open water surface in the "bareheaded" vessel.

Digressing a bit, we would like to mention here that the described phenomenon of the increase of pressure during the heating of liquid in a closed volume has been successfully employed in the design of a utensil called the pressure cooker (familiar maybe to those who still do real cooking, at least once in a while). Instead of the spout though, it has a safety relief valve which opens only if the pressure inside goes over a certain limit; the rest of the time, the vessel remains entirely sealed. As liquid inside gets vaporized and all the steam builds in the cooker, the internal pressure rises to about $1.4 \cdot 10^5\,Pa$, before the relief valve opens, so that the boiling point (going back to our useful Table 20.1) moves up to $T_b^* = 108°\,C$. This allows one to cook food much faster than in a regular pot. However, one should be extremely cautious while opening the pressure cooker after taking it off the stove because when unsealed, the inside pressure drops and the liquid becomes significantly *superheated.* Therefore a certain mass δm, such that $r\,\delta m = c\,M\,(T_b^* - T_b)$, will instantly evaporate and can cause severe scalding. (In this situation, the liquid starts boiling explosively in the whole volume of the pot at once).

By the way, at high elevations (in addition to the typical beautiful scenery), the atmospheric pressure is lower and it becomes quite an undertaking to cook, say, a piece of meat, as water starts boiling at $70°\,C$ (the ambient pressure at the elevation of Everest, for example, is about $3.5 \cdot 10^4\,Pa$). So the pressure cooker is usually a quite welcome part of the climber's equipment. Due to its ability to reach acceptable cooking temperatures, it saves fuel as well. This is another weighty justification to have this massive utility in one's backpack.

But let's go back to our kettles, boiling now at full tilt, still on the stove. It's time to take them off. Take notice that the one with the lid doesn't stop boiling right away, after having been taken from the stove: the steam continues to puff out for some time. What fraction of water does actually evaporate during this ("postheating") boil?

To answer this one, we have to look at the chart in Figure 20.3, representing the dependence of the water temperature on height while the water is boiling (the heat, of course, is being supplied through the bottom). From the picture, we see that the slim bottom layer of about $\Delta H = 0.5\,cm$ thick is quite hot, with the temperature drop across it from $T_{bot} = 110°\,C$ (T_{bot} is the bottom temperature) to $T_i = 100.5°\,C$. The rest of the water stays, according to the graph, at about $100.5°\,C$, yet undergoes another step down of $\Delta T = 0.4°\,C$ when approaching the free surface of the liquid (the graph corresponds to the water level in the kettle being $H = 10\,cm$). Thus, for the amount of heat (beyond the equilibrium) stored in the vessel after its heating has stopped, we can write the following formula:

$$\Delta Q = c\,\rho\,S\,\Delta H \left(\frac{T_{bot} - T_i}{2} \right) + c\,\rho\,S\,H\,\Delta T,$$

where S is the kettle's bottom area (the kettle is considered to have a cylindrical shape). Sure enough, the tea-kettle's base is overheated somehow, but because of the much higher specific heat of water, we can safely disregard the contribution of this effect.

The ΔQ amount of extra heat is then spent evaporating the layer of liquid of thickness δH. The mass δm of such a slice of water can be found from the equation of heat balance:

$$r\,\delta m = \rho\,S\,\delta H\,r = c\,\rho\,S \left[\Delta H \left(\frac{T_{bot} - T_i}{2} \right) + H\,\Delta T \right]$$

consequently, giving us for δH:

$$\frac{\delta H}{H} = \frac{c}{r} \left[\frac{\Delta H}{H} \left(\frac{T_{bot} - T_i}{2} \right) + \Delta T \right] \approx 2 \cdot 10^{-3}.$$

Fig. 20.3: Temperature
near the bottom of a
boiling kettle is much
higher than that of
the mass of water.

So it shows that after we take the kettle from the stove, it still loses, due to continuing boiling, about 0.2 percent of its water content.

The typical time it would take to boil out all water of, say, mass $M = 1\,Kg$ from a kettle provided with heat of $\mathcal{P} = 500\,W$, can be calculated as:

$$\tau = \frac{r\,M}{\mathcal{P}} \approx 5 \cdot 10^3\,s.$$

Therefore, the 0.2 percent of its mass will be vaporized within about 10 s (supposing that the evaporation rate does not change with respect to the stationary regime).

And now, after all this discussion, the time arrives to serve our long-awaited tea. By the way, in the Eastern countries, it's customary to drink tea using tiny tea bowls rather than tea cups or tea glasses. The former were most likely first introduced by the nomadic tribes of Asia – the small bowls are much easier to pack and they are far less fragile, which are obvious conveniences when one has an itinerant lifestyle. Besides, they have one more serious advantage over ordinary glasses: the bowl shape, which is wider at the top, lets water cool down faster in the upper layers precluding a possible burning by the scorching liquid, while the tea in the lower part of the utensil remains hot.

However in Azerbaijan you may meet another type of the tea-drinking vessel, called an Armudi (see Figure 20.4). Here the wide top helps the safe and pleasant cooling of the beverage whereas the spherical part below has a minimum surface area and therefore keeps the tea hot for a longer time. Thus you may enjoy sipping the warm tea as the long and exquisite table-talk goes on.

The nifty porcelain tea-cups (not those over-sized mugs you usually

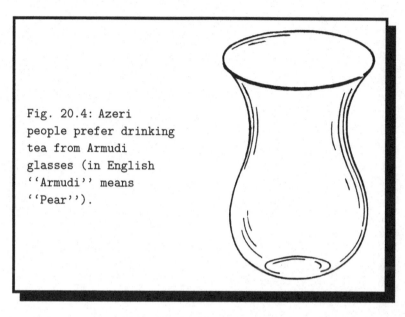

Fig. 20.4: Azeri
people prefer drinking
tea from Armudi
glasses (in English
''Armudi'' means
''Pear'').

get as safety awards from your company), which have been around for centuries, often have this wide-top profile too. The less-advanced cylindrical glasses came into general use as tea-serving ware only in the nineteenth century simply because of their lower cost and were traditionally used by men, whereas the artful china tea-cups were politely left for the better half of the human race. Over time though, the inferiority of the tea-glasses was improved a bit with the invention of glass-holders (back then often decorated with their owner's monograms).

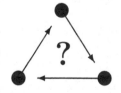

Can you think of what physical criteria material for glass-holders should satisfy? Would, say, aluminum and silver be good candidates?

Chapter 21

The physics of good coffee[a]

> See how easy it can be to make a man happy:
> a nice cup of coffee does it all.
>
> Edoardo De Filippo, *"These Ghosts."*

The traveler who is used to going from country to country has probably observed that, in this century of globalization and multinational monopolies, the very same drinks are being offered in New York or Kathmandu while coffee, on the other hand, is still very much a regional drink. To drink a coffee is a very different experience depending on whether you are in Turkey or Egypt, Italy or France, Finland or America. If you ask for a coffee in a bar in Naples, it will be served in a small elegant cup not much larger than a thimble – at the bottom of which a viscous, nearly black liquid covered with an inviting foam is slowly moving around. On the other hand, if you order a coffee in Chicago, you will get a styrofoam container filled with a half-liter of hot brown water. Here we do not wish to judge which of the two coffees tastes better or makes you feel better; instead, we will simply discuss the various methods employed in preparing coffee, and the relevant physical processes involved in those methods.

---

[a]This chapter was originally published in English (translated from Italian by E. Fortin) under the same title as an article in *Physics in Canada*, Volume 58, No. 1, pp. 13–17 (2002).

21.1   Boiled coffee

This method of preparation is very ancient and has extended nowadays up to Finland and the northern regions of the Scandinavian countries. Roasted coffee is coarsely ground, poured into water (about 10 grams of coffee per 150−190 *ml* of water) and boiled for about 10 minutes. Without filtration, the drink is served directly into cups where it is left to settle for a few minutes. There is no interesting physical process involved here, and the authors will refrain from discussing the taste of such coffee.

21.2   Filter coffee pot

This type of coffee-maker is quite common in America, in northern Europe, in Germany, and in France. Its principle of operation is very simple and the process lasts 6 − 8 minutes. Coarsely ground coffee is poured into a conical paper filter. Boiling water poured from the top "washes" the coffee grains, penetrates across the filter and is collected in a glass container. As a result, one gets a light coffee since only a few of the oils make it through the dense filter. In addition, the coarsely ground coffee granules and the lack of pressure do not favor the complete extraction of all the flavors and aromas present in the coffee. In America, the norm is 5−6 grams of coffee per 150−190 ml of water, while in Europe, it is 10 grams of coffee per cup of water.

21.3   Turkish coffee

We continue our discussion with a process of preparation known as "coffee a la Turk". In this case, the coffee beans are ground to a fine powder and this powder mixed with sugar, is poured into a conical shaped metal container (usually copper or brass) which is called an "ibrick", Fig 21.1. Cold water is then poured into the ibrick and the container is immersed in red hot sand.

   In another recipe, the powdered coffee is put into the container which has already been filled with boiling water (if you do not have easy access to hot sand, you can use a gas burner or electric stove). The heat transported from the sand across the bottom and the sides of the ibrick generates convective currents while heating the liquid: as it moves, the hot liquid brings

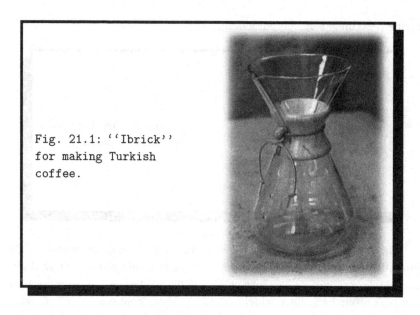

Fig. 21.1: ''Ibrick''
for making Turkish
coffee.

up some of the coffee powder on the container's inner surface. Thanks to surface tension, a "crust of coffee" is formed on the surface. Little by little, the contents of the ibrick reach the boiling point, and bubbles start to break the crust forming a kind of foam. At this point, the process is interrupted by removing the ibrick from the sand so as not to damage the taste of the coffee. This process must be repeated twice more until a thick layer of foam has formed. The contents are then poured into small cups and one waits for the coffee powder to settle on the bottom. At the end, one obtains a dense and tasty beverage (especially if a small quantity of water is used).

21.4 The Italian moka

The "moka" is the most commonly used coffee maker in Italy. It is made of three parts: the base where the water is to be heated, a cylindrical metallic filter where the finely ground coffee powder is deposited, and the top section which has the approximate shape of a truncated cone, where the finished coffee arrives, Figure 21.2. The filter is the heart of the moka: below the filter is a fixed metallic funnel which draws very near the bottom of the base container. The moka does not leave much room for inventiveness in preparation. Instead, in contrast to other methods of preparation, the recipe determined by the pot's design must be followed.

Fig. 21.2: The Italian moka coffee maker

The process of making coffee with the moka is very interesting. The ground coffee is gently pressed into the filter and cold water is poured into the base container. The moka is closed by tightening the base and the top part around the filter which covers the water contained in the base. A rubber ring seal insures the tightness between the two parts. The water is heated on a weak flame and, consequently, the vapor pressure over the surface of the water increases rapidly, forcing the water to rise within the stem of the funnel leading through the coffee powder contained in the filter. The coffee then rises through a thin tube to the upper part of the pot. At this point, the coffee may be served in small cups.

This method appears simple and easy to understand. But what is the moving factor behind the process? Certainly the heat from the flame. Initially, the water is heated in a closed space – where the water itself occupies most of the available volume. The water quickly reaches $100°\,C$ (boiling temperature at sea level, that is 212° Fahrenheit) and the saturated vapor pressure above the water reaches 1 atmosphere. Providing further heat results in an increase in both water temperature and saturated vapor pressure: water and vapor remain in equilibrium while temperature and pressure keep on increasing. (The saturated vapor pressure as a function of the temperature may be found in Table 20.1, page 172). On the other hand, the external pressure (above the filter) remains equal to the atmospheric pressure. The saturated vapor which is at a temperature slightly above $100°\,C$ acts as a spring, pushing the boiling water across the coffee powder. In this way, the extraction of all the flavors and aromas – the essence of the coffee – and other components, transform the water into a tasty beverage.

Obviously, the taste depends on the quality of the coffee powder, the water temperature, and on the time required for the water to cross the filter. The secrets for the preparation of the coffee mixture are those of each producer, based on talent, work, and long experience. What determines the transit time of the water across the filter can be understood without resorting to industrial espionage – relying solely on the basis of physical laws.

In the middle of the nineteenth century, two French engineers, A. Darcy and G. Dupuis, performed the first experiments on the motion of water through sand-filled tubes. This research represents the start of the empirical theoretical development of filtration which is successfully applied today to the motion of liquids through solids having interconnected pores and cracks. It was this Darcy who formulated the so-called law of linear filtration which today bears his name. That law relates the mass of fluid/sec (Q) going through a filter of thickness L and cross section S with the pressure difference ΔP across the filter:

$$Q = \kappa \frac{\rho S}{\eta L} \Delta P, \tag{21.1}$$

where ρ and η respectively represent the density and the viscosity of the fluid, κ, (the constant of filtration) does not depend on the characteristics of the liquid, but only on the porous medium.

In the original work of Darcy, the pressure difference ΔP was assumed to originate exclusively from the force of the weight. In that case, $\Delta P = \rho g \Delta H$, where g is the acceleration of gravity and ΔH represents the height difference between the two extremities of the filter which are supposedly vertically mounted. From Eq. (21.1), κ has the dimension of the area (meter^2). The values of κ are in general very small. For example, in a large grain sandy terrain, the value of κ is of the order $10^{-12}-10^{-13} \, m^2$, and in a terrain of pressed sand $\kappa \sim 10^{-14} \, m^2$,

Let us try now to apply Darcy's law to the study of our moka coffee maker. For example, it would be interesting to know the temperature reached by the boiling water within the base of the moka. This temperature can be obtained from the dependence of the boiling temperature on the pressure and Darcy's law:

$$\Delta P = \frac{m}{\rho S t} \frac{\eta L}{\kappa} \tag{21.2}$$

The usual thickness of the filter in a moka for three persons is $L = 1\ cm$, $S = 30\ cm^2$ and $m = 100\ g$ of coffee is prepared in $t = 1$ minute. For the coefficient of filtration, we can use that of a terrain of coarse sand $\kappa \approx 10^{-13}\ m^2$, $\rho = 10^3\ Kg/m^3$. As for the viscosity, one must be careful because it is temperature dependent. In the tables of physical quantities, $\eta(100°\ C) = 10^{-3}\ Pa \cdot s$ from which $\Delta P \approx 4 \cdot 10^4\ Pa$ follows. From Table 20.1 one sees that the corresponding water boiling temperature is $T^* \sim 110°\ C$ $(230°\ F)$.

We have thus understood the physical basis of the coffee preparation process in an Italian moka. However, there are obscure aspects in that machine which can transform it into a real bomb, a menace for the kitchen walls and ceiling, and of course for the safety of whoever may be in the vicinity. Why and how can it happen? Clearly, such a disaster can only occur if the safety valve located in the base of the coffee maker is obstructed (old coffee makers are dangerous!), and at the same time, if the normal flow across the coffee filter is blocked off. For example, too fine a coffee powder compressed too hard could become impenetrable for the water. If both of these conditions occur, the pressure in the base can increase to a point where the thread upon which base and top are screwed breaks down.

Let us now try to explain what happens when the coffee powder is packed too tightly. We must remember that Darcy's law is empirical in nature. At the microscopic level however, the filter can be imagined as a system of interconnected capillaries of various sections and lengths. The validity of Darcy's law requires a laminar flow of the liquid across the filter. In reality, various irregularities in the capillary structure result in the formation of vortices and consequently a transition to the turbulent mode. These dissipative phenomena require an increase in pressure in order to keep a constant flux across the filter. Another factor that determines the lower limit for the transparency of the filter is related to the surface tension. In fact, it is known that in an ideal capillary of section r, one must apply a pressure difference $\Delta P > 2\sigma/r$. Taking the capillaries' radius as $\sim 10^{-4}\ m$ and remembering that σ is about $0.07\ N/m$, one obtains a threshold of $10^3\ Pa$ for the filter's transparency. This value is about one order of magnitude lower than the pressure difference across the filter at $110°\ C$. It is clear that, if too fine a coffee powder is packed too hard within the filter, the effective radius of the capillaries could become much smaller than the $10^{-4}\ m$ previously estimated. Under these conditions the filter could, in practice, turn "opaque" or impenetrable.

Coffee pots transformed into bombs can indeed be quite dangerous. Let us start from the worst situation: the filter and safety valve are blocked and 100 grams of water are heated in a volume not much greater than that occupied by the water. For temperatures near the water critical phase (where the density of the vapor and of the water are comparable), at a temperature of $T = 374° C = 647 K$ ($T = 705° F$), all of the water becomes vapor. In principle it is possible to heat the moka further, but at higher temperatures, the pot will start to glow (this has not been observed yet)! Thus, for a reasonable calculation one can assume a final temperature of $T = 600 K$ ($T = 620° F$). The vapor pressure in the base of the moka can be easily estimated using the ideal gas equation:

$$PV = \frac{m}{\mu} RT. \tag{21.3}$$

Since $m = 100\,g$, $V = 120\,cm^3$, $\mu = 18\,g/mole$, $R = 8.31\,J/(mole \cdot K)$ we can find $P \approx 10^8\,Pa = 10^3$ atmospheres. The associated energy is impressive, reaching

$$E = \frac{5}{2} PV \sim 50\,KJ \tag{21.4}$$

and the explosion could thus throw moka parts at velocities of up to hundreds of meters per second. From that calculation, it is obvious that the explosion must occur much before the temperature of $600\,K$ is reached. But it demonstrates the strong forces developed within the moka with excessive heating: forces sufficient not only to dirty up the kitchen with coffee but also to be the source of other problems. A curious effect related to the explosion is the following: if you are near the moka when it explodes, and you are lucky enough not to be hit by pieces of metal traveling at high velocities, you will, in all probability, be hit with overheated vapor jets and coffee powder. Your first reaction will be to quickly remove your wet clothes to avoid being scalded. A moment later, you may realize, with great surprise, that you have a feeling of coldness rather than heat.

The explanation for this surprising effect is rather simple: the vapor expansion following the explosion is very rapid, not allowing time for a heat exchange with the ambient. In other words, the vapor expansion is adiabatic and with good approximation obeys the related ideal gas laws:

$$TP^{\frac{1-\gamma}{\gamma}} = const, \tag{21.5}$$

where $\gamma = \frac{C_P}{C_V}$. Here C_P and C_V are the molar heat capacities of water at fixed pressure and fixed volume respectively.

As is well known from molecular physics, C_P and C_V can be expressed by means of the number of the degrees of freedom i. Indeed, as we know from the law of the equal distribution of energy, for each degree of freedom at fixed volume there is a corresponding contribution to heat capacity equal $R/2$, while $C_P = C_V + R$. For a molecule of water $i = 6$ and $\gamma = \frac{i+2}{i} = 4/3$, hence

$$T \sim \sqrt[4]{P}. \tag{21.6}$$

Assuming the initial temperature and pressure of the vapor to be $500\,K$ ($440°\,F$) and 10 *atm* respectively, one sees that, upon reaching a pressure of one atmosphere, the gas may attain a temperature as low as $100\,K$ ($-280°\,F$).

In conclusion, coffee prepared in a moka is strong and aromatic but does not reach the quality of an espresso served in a good bar. The main reason is the high temperature of the water forced through the filter by the vapor. Here, then, is some advice to improve the quality of coffee prepared with the moka: warm the coffee pot slowly; in this way the water will traverse the filter slowly and, at the same time, excessive heating of the vapor in the base container will be prevented. Finally, a good coffee can be prepared with a moka in an alpine refuge where the external pressure is less than one atmosphere (for example, at an altitude comparable to that of Mount Everest, the water boils at about $70°\,C$ ($\approx 160°\,F$)) and the superheated water reaches only a temperature of $85 - 90°\,C$ (the best for coffee preparation).

21.5 The ancient coffee maker – "la Napoletana"

This coffee maker reminds us somewhat of the moka with the main difference being that the "motor" forcing the liquid through the filter is gravity rather than vapor pressure. The coffee pot consists of two containers, one above the other and separated by a filter filled with coffee, Figure 21.3. When the water from the lower container reaches the boiling point, the coffee pot is inverted. The filtration then occurs under the action of the pressure of a water column of a few centimeters (ΔP on the filter does not reach $10^3\,Pa$). The coffee preparation process is slower here than for the moka. It is interesting to make an experiment and prepare the coffee in the two different pots and verify using Darcy's law where the preparation time is inversely proportional to the pressure, the validity of which was stated above. According to the connoisseurs, a coffee prepared "alla Napoletana" is better than that made in a moka: here the filtration process occurs more slowly and the flavor of the coffee is not damaged by contact with super-heated water. Unfortunately, the pace of modern life does not leave us time for calm discussions while waiting for the coffee "alla Napoletana". This luxury remains only in the paintings showing the Neapolitan life of the past and in the works of Edoardo di Filippo.

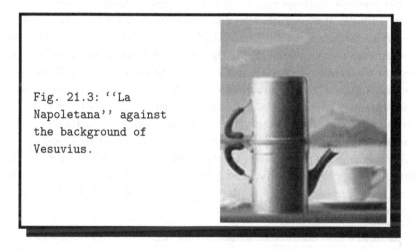

Fig. 21.3: ''La Napoletana'' against the background of Vesuvius.

21.6   The "Espresso"

Not all Neapolitans were so patient however. It seems that in the past century, a resident of the Kingdom of the Two Sicilies, who could not wait patiently for the coffee "alla Napoletana", managed to convince an engineer friend from Milan to devise a new type of coffee maker that could prepare a good, dense, aromatic coffee in less than half a minute. Nowadays, a cup of good coffee involves a treasure box of secrets about the growth and harvesting of coffee beans, the preparation of the mixture, and, obviously, the roasting and grinding of the beans. Behind the art of coffee making stands a sophisticated technology: the espresso machines are much more complicated than the simple coffee makers described earlier.

Normally, espresso machines are found only in bars and restaurants but, for enthusiastic coffee lovers, there also exists a home version of the machine, Figure 21.4. In the professional type of machine the water, at a temperature of 90–94° C, is pushed at a pressure of $9-16$ atmospheres across the filter containing finely ground coffee. The whole process lasts $15-30$ seconds and provides one or two small cups ($20-30$ ml) of espresso. The mechanism responsible for the passage of the liquid across the filter can also be explained by Darcy's law as for the moka. However, the pressure here is 10 times greater while the temperature is around 90° C. A higher pressure increases the speed of travel across the coffee powder while a lower temperature ensures that the unstable components contributing to the coffee flavor are not decomposed. Although this must appear strange, the espresso contains less caffeine than the coffee prepared with the moka because the contact between water and coffee powder is shorter (20–30 seconds as compared to 4–5 minutes for the moka). The first espresso machine was presented in Paris in 1855. In the modern machines used in bars and restaurants, the water reaches the required pressure with the help of a pump. In the classical machines, one first raises the lever to introduce the water contained in the heating cylinder situated within the vessel under the filter. Successively, the same lever is lowered to push the water across the filter. The pressure applied to the liquid is thus provided by the force of the arm multiplied by the lever action.

Fig. 21.4: Even the house version of the espresso machine looks impressive.

It is interesting to observe the fall of the coffee as it exits the spout of the machine. First, the liquid exits in a continuous jet, becoming progressively weaker, and eventually ending up in drops. Let us try to understand this phenomenon with a simple analogy: the authors have observed the same effect in the mountains when the sun warms up and loosens the snow on a roof. Again in this case, the water can fall from icicles as drops or as a continuous jet. Let us try to estimate the critical flux for this transition.

Let us assume that the water falls slowly. It is clear that when the water flow is too small, no jet occurs. At the bottom of the icicle the water forms a drop, the drop grows slowly, reaches a critical dimension, falls, and the process repeats itself [b]. As long as the flow is low, the process is quasi static. In equilibrium conditions, the drop is detached when its weight mg exceeds the surface tension force

$$F_\sigma = 2\pi\sigma r, \tag{21.7}$$

where r is the radius of the drop's neck:

$$mg = 2\pi\sigma r. \tag{21.8}$$

The time required to build a drop is:

$$t_d = \frac{m}{\rho Q_d}, \tag{21.9}$$

where Q_d is the volume flux and ρ the density of the liquid.

Under the combined action of the surface tension and gravity, the drop is in equilibrium. When its mass reaches a critical value and the surface tension force cannot compensate for the weight, the bond between the icicle and the drop breaks down. The related characteristic time τ can be calculated using the theory of dimensionality: liquid with viscosity η under the effect of the surface tension with the coefficient σ has to be displaced on the distance of the order of r. In the present case it is necessary to write a dimensional relation that links the breakdown time τ with the characteristic length of our problem, i.e. the radius r. Exactly the same problem was already analyzed in Chapter 11 and the result (see (11.2)) reads as:

$$\tau = \frac{r\eta}{\sigma}. \tag{21.10}$$

---

[b]The process is shown in detail in Figure 10.8, page 85.

Clearly, the change from the drop to the continuous jet regime occurs when a drop being formed is joined by another before the fall of the first one, i.e.

$$t_d \sim \tau \qquad \text{and} \qquad \frac{m}{\rho Q_d} = \frac{r\eta}{\sigma}. \tag{21.11}$$

Extracting the drop's mass from the equilibrium condition (21.11) we finally get a clear expression:

$$Q_d \sim \frac{2\pi\sigma^2}{\eta\rho g}. \tag{21.12}$$

We see that Q_d is independent of the dimensions of the icicle's tip. In the case of the espresso machine, the area of the spout can actually influence the value of the critical flow. However, according to equation (21.12), the critical flow cannot show a strong dependence on the spout's cross section.

21.7   Instant coffee

The pressures of present-day living have created this type of coffee. It is prepared from a real coffee which is evaporated at high temperature and low pressure[c]. The resulting powder is preserved in a vacuum container where it can keep for a long time. To prepare the drink, the powder is simply thrown into hot water.

21.8   Variations on the Espresso theme

With an espresso maker and a good coffee mix, it is possible to be inventive. For example, in Italian bars, you can drink: a concentrated *"caffè ristretto"* ("restricted" coffee) which is prepared with a standard quantity of coffee but with less water; a *"caffè lungo"* ("long" coffee) prepared with a normal quantity of coffee but with more water; a *"caffè macchiato"* ("dappled" coffee) which is espresso with a dash of milk; a *"caffè corretto"* ("correct" coffee) which is espresso with a liquor; a cappuccino which is espresso in a medium sized cup to which is added milk "beaten" with the vapor in order to obtain a light, frothy foam. A good barman can pour the milk on the

---

[c]Dehydration can be done by spray drying in hot air, drying under vacuum, or by freeze drying. The methods vary among manufacturers. —A. A.

coffee so as to write the first letter of your name on the surface. Finally, one can simply add a bit of cacao to the foam.

They say that nowadays in Naples, a few bars still serve the *"caffè prepagato"* ("prepaid coffee"). A well-dressed gentleman accompanied by a lady enters a bar and orders three coffees, two for themselves, and one "prepaid". After a short time, a poor man enters the same bar and asks, "Is there a *caffè prepagato?*" And then the barman pours him a cup of free coffee.

Naples always remains Naples... and not only in the Toto movies.

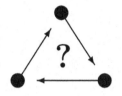

Why is the time of drop detachment from an icicle the same as the time of drop coalescence from Chapter 11?

Chapter 22

"Nunc est bibendum": physicists talk around glasses of wine

Oh, beloved child, bring the precious chalices,
for the son of Zeus and Semele
gave to human being the wine
to let them to forget their sorrows

Alcaeus[a].

Wine is one of the most civilized things in the world
and one of the natural things of the world that has been
brought to the greatest perfection, and it offers a greater
range for enjoyment and appreciation than, possibly, any
other purely sensory thing which may be purchased.

Ernest Hemingway[b], *"Death in the Afternoon."*

22.1   On the origin of wine and the methods of wine making

What can we add to the idea of wine created over the centuries by poets, writers, journalists and enologists? The epigraphs to this chapter were written over a 2,500 year period: the first one comes from one of the earliest poets of our civilization, the second from one of the most popular modern writers. These remarks praise wine and are the manifestation of the love of land and for one of its most valuable fruits, transformed by labor and skill

[a] Alcaeus (620–580 B.C.), Greek poet.

[b] Ernest Hemingway (1899–1961), American writer.

197

into a divine drink, playing a special role in the life of human beings. These remarks are also somewhat related to what we are going to talk about later on.

The origin of wine can be traced back to prehistoric caves, where archeologists have found records of grapes. Thousands of years B.C., the fruits of the grapevine, which presumably originated in India, were known. According to a legend which would be hard to verify, wine was discovered at the court of a Persian king. A distressed lady living in the royal palace decided to kill herself. Following the suggestion of a priest-like man appointed to perform religious rites she drank the strange-looking liquid being formed at the bottom of large bowls filled by grapes. The result was unexpected: her depression was gone, gloomy thoughts gave way to happiness. This experiment highlighted the felicitous qualities of the drink and, after testing it personally, the king introduced the wine for regular consumption. Since that moment, the status of wine had been on the ascent and it also became a sacramental drink. It is well known that in the Christian tradition, wine is part of the ritual for the Holy Communion. In the Jewish ritual of Kiddush, the ceremony of blessing is performed with wine and bread. Arabic legends are claimed to state that Adam was familiar with the vines, and it was the grape rather than the apple that was the "forbidden fruit". The migration of Aryans from India to the west around 2500 B.C. led to a spread of these methods throughout the ancient world. It is certain that wine was being produced in Sicily about 2000 B.C., probably due to colonization by Greeks and Egyptians. Some time later, Sabinis, who lived on the territory of modern Italy (recall the painting "The Rape of the Sabines" by N. Poussin) and Etruscans, who populated modern Tuscany 3000 years ago, started making wine. It is known that in ancient times, people acknowledged the healing properties of wine, and also used it as an antiseptic remedy for treating wounds. Wine-based medical prescriptions were found in Egyptian papyruses. Hippocrates prescribed wine as an anti-inflammatory remedy. Speaking of Romans, we may remember how Horatio and Virgil eulogized wine in many of their compositions. In 42 B.C., Columella wrote an excellent instruction book on viticulture. In the first century A.D., winemaking was widely spread throughout the Roman Empire. In his struggle against overproduction, Emperor Domician prohibited an increase of the area of grape cultivation in the Roman Empire and ordered that the vineyards in the provinces be cut by half. By 200 A.D., the crisis in agriculture related to the Barbarian invasions led to the decline of viticulture, and soon it completely disappeared due to the advent of Muslims in the historic arena, for the use of alcohol was prohibited by the Koran.

In the Middle Ages, wine was sporadically produced in castles or in monastic minsters. Only the period of the Renaissance i.e. beginning in the sixteenth century) in Europe restored the mass production of wine. As Hemingway said, the progress in wine making, along with the revival of arts and culture, highlights the development of civilization and liberation from barbarism. Modern enology – the science of wine making – originated at the beginning of the twentieth century, due to dramatic events in viticulture: the mass migration of people caused the transfer of unusual parasites into traditional wine-making areas, which caused the destruction of vines. Some of these parasites such as powdery mildew (*Uncinula necator*) and downy mildew (*Plasmopora viticola*) were repelled by chemicals based on sulphur and copper. A notorious phylloxera (*Phillossera vastatrix*), however, proved to be resistant to pesticides. To get rid of it, enologists turned to genetic methods. The grape vine brought to America by European settlers long ago was gradually evolving in the new environment. Its roots had developed an external protective tissue, which made it immune to phylloxera. Therefore, to save the best European plants, the *Vitis vinifera*, enologists began grafting onto them well-weathered species native to the United States (*Vitis riparia, Vitis rupestris*, and *Vitis berlandieri*). Their offsprings are literally at the root of modern European viticulture. Old European vines based on *Vitis vinifera* survive only in several small areas in Europe: in Jerez (Spain) and in Colares (near Lisbon), where sandy soil prevents the growth of phylloxera, and also in Moselle or in Douro areas, where shale soils play an analogous role. Today the battle against diseases of the vines is a primary concern for the enologists, though sometimes parasites can be useful. For example, the gray mold *Botrytis* is regarded as a beneficial kind of mold: in Sauternes frequent fogs in the fall encourage the proliferation of this infection. In the most infected berries, the mold absorbs up to 20 percent of the moisture, thus increasing sugar content in the grape juice and creating a unique taste in Sauternes wines[c], delicious in combination with Roquefort[d] and pâté de foie gras[e].

[c]Sauternes (France).

[d]Roquefort – famous French cheese with lots of green and blue mold.

[e]Pâté de foie gras – paté made of the enlarged liver of a specially fed goose.

After the grapes are harvested and pressed the process of fermentation begins. Wine berries have a special property: it is enough to scratch their skin and leave them for a short time in grape juice and a natural process of fermentation – conversion of sugar into alcohol – will begin. This process takes place due to the presence of *Saccharomyces ellipsoideus*, the yeast derived from the grape skin. Sometimes yeast cultures are added to the grape juice; this method is called "induced fermentation". During the fermentation process, sugar is converted into ethyl alcohol, carbon dioxide, glycerin, acetic and lactic acids, and many other substances. We mention here only the substances essential to our pending story. The attempt to describe in detail how to create good wine would be as hard as the attempt to describe in detail how to paint a famous masterpiece of art or to provide instructions on carving a masterpiece from a marble slab. Here we will confine ourselves to the description of certain physical phenomena occurring in wine making. Grape skin contains yeast cultures that convert juice into both wine and vinegar; when left unguided the yeast cultures can spoil the beverage. Thus an enologist once said that the Lord had not been planning to transform grape juice into wine: vinegar should have come out in the end. To avoid such an undesirable outcome and to streamline and harness the fermentation process is the responsibility of man. Modern methods of fermentation based on the knowledge of chemical processes and high technology offer better results than home wine-making. Today, for example, to produce high-quality wines enologists turn to limited areal density of plants yielding better and more precious grapes. Earlier, wine makers preferred planting tall plants, as they were more convenient for harvesting. Now, to improve the quality of the wine, short plants are grown, with berries getting extra heat from the sun-warmed soil. In the past, the adverse effect of acetic yeast was controlled by certain chemical components. Today, alternative methods based on fine filtration and deep refrigeration are frequently used.

Fermentation is an exothermal process that is accompanied by a marked generation of heat. As a result, uncontrolled production of wine may lead to a temperature rise in must to temperatures as high as $40-42°\,C$. In this case, the exquisite and volatile flavors of fruits and flowers that could create a unique bouquet are lost. In modern production, the fermentation is performed in double-walled stainless steel tanks reminiscent of Thermos flasks. Coolant circulating between the walls keeps the temperature of must low – no more than $18°\,C$, thus helping to preserve the wine flavors. However, this method requires a longer time for fermentation; it takes up to three weeks instead of $7-8$ days to complete fermentation at low temperature under natural conditions.

Fine filtration and cooling also help to clear wine of certain sulphites which had been added previously to slow down the acidification. Two factors play a role in this process. Each additional percent of alcohol lowers freezing temperature for wine by $0.5° C$ relative to $0° C$. Thus, wine which contains 12% of alcohol will freeze at about $-6° C$. At the same time, sulphites form organic complexes with the peptides found in wine. These complexes form small crystals that can be filtered off even before wine is frozen. A quick view of the previous methods explains why a contemporary winery more and more resembles a research laboratory. Low temperatures also help wine makers increase relative sugar content, so grapes are slightly frozen before pressing. Some water is condensed into ice and the sugar content is higher in the residual juice (the so-called *ice-wine* in Germany, Austria, and Canada). Another way to increase sugar concentration is to dry grapes (Erbaluce, Caluso) for sweet wine manufacturing: the so-called passito (Sicily and other regions of Italy), muscat (Crimea) and other wines. Let us now explore what we regard as several interesting physical phenomena related to alcoholic beverages and their consumption.

22.2 Tears of wine

If we swirl wine in a glass, we can observe an interesting phenomenon: it often leaves a kind of coating on the inside of the glass, that separates into viscous-looking rivulets, called legs or tears. These legs slowly slide down the glass, returning to the wine's surface. Sometimes it is believed that the presence of the tears indicates the high quality of the wine. Connoisseurs of wine like to discuss the tears at the table using terms like glycerin and corpulence (full-body).

Let us analyze the physical origin of this phenomenon. First of all, let us note that in order to inspect "wine tears" in a strong enough alcohol-water solution (more than 20% of alcohol) there is no need to swirl or shake the glass. The unusual effect of mass convection up the wall in the form of a thin film and its counter flow down as "wine tears" is observed, in this case, even in a standing glass. In hydrodynamics the first part of this phenomenon is known as the Marangoni effect. It consists of the motion of a film boundary in the direction opposite to the gravitational force due to the appearance of a surface tension gradient over the height of such a film. The point is that alcohol vaporizes from the film's surface faster than water. The surface tension of water is larger than that of alcohol. Consequently, nonhomogeneous vaporization of the alcohol from the surface of films of varying thickness (along the wine surface in glass) results in the appearance of an alcohol concentration gradient and hence to the gradient of the surface tension. This inhomogeneity leads to the generation of a force pulling the film up the glass wall. The details of such boundary motion were studied in 1992 in the experiments conducted by French physicists Fournier and Cazabat[f]. They found a surprising relation between the film boundary displacement (climb) L and time t:

$$L(t) = \sqrt{D(\varphi)t}.$$

In fact, such functional dependence was already well known in physics. It was first derived theoretically by Einstein and Smoluchovski, and it established the relation between the effective particle displacement in its diffusive motion and the diffusion time. The coefficient D in the Einstein-Smoluchovski formula is a so-called diffusion coefficient and it is defined by means of the particle velocity and its mean free path. In their experiments, the French scientists also discovered the nontrivial dependence of their diffusion coefficient $D(\varphi)$ on the solution's alcohol content (proof) φ (see Figure 22.1).

As was already mentioned, when the alcohol concentration in solution was less than 20% ($\varphi < 0.2$), very little spontaneous film climb up the glass wall was observed. Strikingly, in contrast to intuitive expectations, the largest values of the coefficient D were obtained for the relatively dilute solutions, even though the alcohol evaporation is going slower. Looking at the experimental data (Figure 22.1), one may notice that the diffusion coefficient D reaches the minimum at $\varphi = 0.9$ and surprisingly does not disappear even when $\varphi = 1.0$, i.e. even pure alcohol climbs up the glass wall. The authors attributed this phenomenon to the absorption of water vapor from the atmosphere by the pure alcohol, which results in the appearance of a small surface tension coefficient gradient. Their assumption was verified by repeating the same measurements for pure alcohol in a dry atmosphere. The corresponding coefficient D was found to be smaller and monotonically decreasing with the growth of φ, and is reduced to zero at $\varphi < 0.2$ (see

---

[f]J. B. Fournier, A. M. Cazabat, "Tears of Wine", *Europhysics Letters*, 20, 517 (1992).

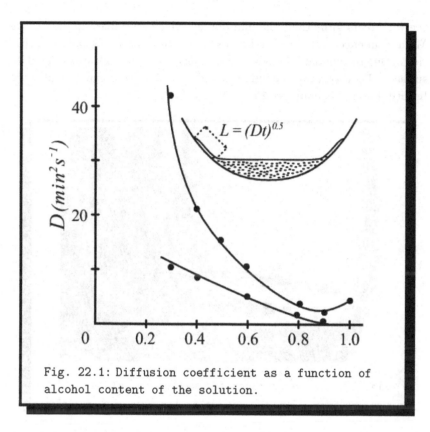

Fig. 22.1: Diffusion coefficient as a function of alcohol content of the solution.

Figure 22.1).

Now let us move to a discussion of the wine tear formation. The first and most evident reason for the formation is the force of gravity, which counteracts the surface tension forces and tends to bring the runaway liquid back. Nevertheless, the so-called Raleigh instability is responsible for the nonhomogeneous character of this return (in the form of separate rivulets). The insight into this complex phenomenon can be gained through examples. When a boat floats onto a lake it generates a wave perturbation of the water surface. As time goes by, these waves decay and the water surface returns to its initial state. One can demonstrate that if the lake had been turned upside-down i.e. the gravitational force had been directed bottom-to-surface rather than surface-to-bottom, the smallest deviation of the surface from the horizontal line would grow up to a squall and destroy the boat instead of decaying peacefully. This instability is responsible for the wine tears formation: the film border, which scrambles up due to the Marangoni effect, turns out to be unstable with respect to small perturbations. Due to random inhomogeneities, small droplets appear on the glass surface. They go down engulfing liquid from border domains and leave beautiful arches behind (see Figure 22.2).

Fig. 22.2: Wine tears on the walls of a glass.

Hence, the more alcohol the solution contains, the less pronounced the Marangoni effect turns out to be, the weaker the binding effect of the surface tension on the wine tears formation, the more wine weeps appears and the less distance between the streamlets is to be observed. Nevertheless, this dependence turns out to be so weak that it does not allow the use of wine tears as a reliable indicator of a wine's proof without looking at the bottle label. With regard to glycerol ($C_3 H_8 OH$), the content of this sweet alcohol in wine is normally very low (it is of the order of 1–2%) and despite the fact that it certainly affects the taste of wine, it has almost no impact on its proof. Regular ethyl alcohol determines wine's characteristics and contents. It should be mentioned that low ethyl alcohol wines are less viscous, hence it seems that the glycerol content does not even determine the wine viscosity. Thus, the observation of the wine tears or even their flow does not provide grounds for serious judgement on the glycerol content in wine.

22.3 Champagne and its bubbles

Strictly speaking, only a sparkling wine produced by a special process in the French province of Champagne has the legal right to carry the name Champagne. The rest are just different sparkling wines, and they are many. Sparkling wine is produced by a special type of fermentation; it can be made from any grape variety, although, as a rule, only three varieties are used: Pinot Noir, Pinot Menier, and Chardonnay. Of these grapes, only Chardonnay is actually white. If the label indicates Blanc de Blanc (white from the white), it means that the sparkling wine was manufactured from pure Chardonnay. Blanc de Noir (white from the black) means that Chardonnay was not used for this wine. There is also Champagne rosè. There are two ways of creating Champagne rosè: either to "correct" white wine with dry red wine, or to keep the unfiltered juice together with skin for a certain period of time, then to filter it off and proceed with making Champagne.

22.3.1 *Méthode Champenoise (Champagne method)*

If you see these words on the label, it means that the wine was produced by the following method: in Champagne, grapes are harvested manually to avoid damaging the berries, thus the yeast from the grape skin will not get into the juice. Grapes are pressed twice. After the first pressing, 80 percent of the juice is extracted, and Champagne produced from it will be called Cuvee. Champagne produced from the remaining 20 percent will be named Taille; the best houses of Champagne do not make it, they only produce Cuvee. There are several ways of saturating wine with carbon dioxide. The method we are dealing with begins with white wine fermentation as above. Then the wine is combined with reserved wines from prior years and bottled[g]. Later some sugar and yeast are added into every bottle and the bottle is closed with a temporary cork resembling a beer cap. The process of secondary fermentation begins. Champagne ages in the bottle for several years: usually up to three, although for special occasions it may stay there for up to six years. The law says that Champagne has to age in a bottle for no less than a year. The next step is to remove the yeast from Champagne. In this process, the bottles are placed into special racks where

[g]A mixture of new and old wines is called dosage. Depending on the amount of added sugar, Champagne will vary in dryness or sweetness and will be labeled Brut (dry), Extra Sec (extra dry), Sec (medium sweet), Demi-Sec (sweet), Doux (very sweet).

they are turned and inclined at 15° per day so that the dead yeast settles on the cork. The last operation is the evacuation of the yeast deposits and the replacement of the temporary cork with a permanent one. This process is not easy, because by that time Champagne is supersaturated with carbon dioxide. It takes a real virtuoso of *Méthode Champenoise* to open the bottle without shaking it and to replace the temporary cork with the yeast deposits on it with a permanent cork. Modern cryogenic technology has streamlined production: the neck of the bottle is frozen, the deposit is removed, and the bottle is corked again.

22.3.2   Bubbles and foam

We all know that sparkling wine is sold in special bottles that can resist high pressure. The reason is that a large amount of carbon dioxide in a metastable state can be released if the external conditions change. Thus, for instance, there are different ways to open a bottle of Champagne, depending on the purpose, skill, and temper of the person who is trying to open it. First, you can slightly let the gas out of the bottle, slowly pull the cork out, and pour the drink into the flutes without spilling a drop; or you can let the cork hit the ceiling to mark a New Year and spill half of the bottle as foam. That is how Michael Schumacher, Formula-1 winner, celebrated his victory. The secret is simple – to get a lot of foam you should vigorously shake the bottle before opening it. As a result, the amount of gas that is present in the neck of the bottle will mix with Champagne and form numerous bubbles. When the cork shoots up, the pressure inside the bottle sharply falls down and the bubbles formed earlier serve as the centers of emission of the diluted carbon dioxide in all the volume of the liquid. Thus, a bottle of Champagne serves as a fire extinguisher and shows the emotions of the Formula-1 winner. Champagne bubbles may entertain not only the viewers of Formula-1. Undergraduate physics majors may ignite the curiosity of language majors of the opposite sex by casting a chunk of chocolate into a glass of Champagne or mineral water. The reader himself can try it out and see that the chocolate chunk will go down, then resurface some time later, and then submerge again. These oscillations will take place several times. We are leaving this little riddle for the reader to explain and move forward to discuss the other less obvious properties of Champagne bubbles.

Speaking about Champagne, physicists cannot overlook its unusual acoustic properties. In Chapter 9, we already touched upon this topic, explaining how Champagne bubbles reduce the crystal tingle of glasses. Here we will concentrate on the other phenomenon: particularly "perlage" i.e. hiss of the bubbles bursting at the Champagne surface. The fizzing sound of the fresh, just-poured-in-glass, Champagne is caused by the emergence of miniature avalanches in its foam. The hissing sound is the sum of multiple separate spontaneous bubble shots. The specifics of such micro-explosions were recently studied by Belgian physicists[h]. If the number of such shots per second had been constant we would have heard "white noise", like the one produced by untuned radio. This kind of a hiss occurs in acoustic processes when the uniform spectrum contains a broad range of frequencies. Careful examination showed that the acoustic spectrum of the bursting Champagne bubbles has nothing to do with white noise. The measurements performed by a sensitive microphone showed that the signal intensity depends considerably on the frequency. This dependence developed due to the fact that the bubbles burst, not independently, but in a collective way, affecting each other. Every shot lasts of the order of 10^{-3} second. Some of them occur in fast consecution one by one and their noise runs into a unique, audible, sound signal. Some bubbles burst independently and they can hardly be heard. The foam blows off as the liquid in the bubble walls falls down under the action of the gravitational force, making the walls too thin for the bubble to survive. Instead of Champagne (whose foam blows off too quickly) scientists studied this process in soapy water, where the foam dissolves very slowly. Such a convenient object allowed them to study the time dependence of the length of intervals between two subsequent bubble shots. They discovered this dependence to be of the power character. This means that the characteristic scale of such an interval just does not exist: the time between two such subsequent events can be an arbitrary – from milliseconds to several seconds – and there is no way to predict the duration of the interval obtained as a result of this measurement. Similar power laws turn out to be characteristic for many natural phenomena like earthquakes (which have no characteristic amplitude), mud flows, and solar flares. Usually they take place in systems where the interaction between the elements plays an important role in the behavior of the system as a whole. Thus, a small pebble falling down a slope that is prone to avalanche can either provoke a terrible disaster or just simply roll down.

[h]N. Vandewalle, J. F. Lentz, S. Dorbolo and F. Brisbois, "Avalanches of Popping Bubbles in Collapsing Foam", *Phys. Rev. Lett.* **86**, 179 (2001).

22.4   "Bread" wine – vodka

Grape vines do not grow well in the North. There, wine is replaced by other beverages produced by distillation of the must of apples (e.g., Calvados in Northern France), plums and apricots (e.g., Slivovica brandy in Bulgaria and the Czech Republic), and, finally, of different sorts of grain (e.g., whisky in the United Kingdom and vodka in Russia). In old times in Russia, vodka was called Bread Wine N21 (Smirnoff). Now it is consumed at dinner in the same way as grape wine in France and Italy. Have you ever considered why all the previous beverages have about 40% alcohol content (ethyl alcohol, $CH_3 CH_2 OH$)? The first reason can be found in a legend based on physical evidence. Russian tsar Ivan the Terrible (1546–1584) introduced a government monopoly on vodka production, and thus made the government responsible for its quality. To increase their profits saloon keepers began diluting vodka, which got them into hot water with their customers who were unhappy with the low quality of their favorite beverage. To put an end to these practices, Peter I issued a decree which allowed the patrons "to beat the saloon keeper to death" if the surface of the vodka he was serving could not be set alight. It was found that 40% alcohol content is the threshold when alcohol vapor can still burn near the surface at room temperature. Naturally, saloon keepers used this lowest level of alcohol content thus reaching a compromise between the lure of profit and concern for their health. There is one more reason for this "magic number". Volume expansion is a well-known phenomenon: when temperature rises, the volume of the objects increases, and when temperature goes down, it decreases, respectively. Most substances (including alcohol) obey this law. Water is different: below $4°C$ its volume starts growing as temperature declines; when freezing, its specific volume jumps up 10 percent! That is why bottled water should not be left outside at temperatures below freezing: frozen water will break the bottles but vodka can be left outside. In Siberia, boxes of vodka bottles are left outside and the bottles stay intact. The explanation consists of two points. First, the presence of such a noticeable alcohol content in the water hinders its crystallization, and thus its specific volume does not jump as the temperature decreases. Second, at volume ratio of 4:6, the total coefficient of volume expansion is close to zero: water "deviation" is compensated by alcohol's "normal" behavior. It is enough to take a look at the tables in the handbook and to compare the coefficients of thermal expansions of water ($\alpha_{H_2O} = -0.7 \cdot 10^{-3}\,\text{grad}^{-1}$) and alcohol

$(\alpha_{C_2H_5OH} = 0.4 \cdot 10^{-3} \, \text{grad}^{-1})$.

The third reason for using 40% was a discovery attributed to the well-known Russian chemist Dmitrii Mendeleev. He is believed to have proved that 40% vodka does not change its alcohol content when left in the open. Thus, a shot of vodka left overnight on the table will be found in the morning filled (unless having evaporated) with the same beverage (which will not happen in the case of wine or champagne) and may help to ease a hangover headache. A curious reader could try to use the results shown in Chapter 20, where the mass loss rate from the liquid surface was estimated for alcohol and water molecules. The result of this calculation gives 55% as the concentration of alcohol for a stable solution, which considerably exceeds the value proposed by Mendeleev and successfully used in production. The point is that the molecules of alcohol and water interact in solution and therefore they cannot be considered in the "gas model" as two non-interacting and interpenetrating liquids. The fourth reason for vodka being 40% proof is related to the viscosity jump which takes place in the vicinity of this concentration in alcohol-water solution. In the dilute solution, alcohol molecules share the environment with the water molecules but practically do not interact with them. Around a concentration of 40% the solution is so dense that they begin to form long one-dimensional chains, i.e., the process of polymerization begins. As a consequence, the viscosity changes drastically and the solution improves its organoleptic properties. The last reason known to the authors is related to the calibration procedure, i.e., to the beverage quality control in the process of home wine-making. When a piece of lard is placed into alcohol-water solution, it will be in the state of indifferent equilibrium only at 40% content of alcohol. In a stronger beverage it will sink, in a weaker solution, it will float on the surface[i].

---

[i]The unmentioned motive that is also attributed to D. I. Mendeleev is it gives the best hydration of alcohol. It is known that, as a rule, the volume of a solution is less than the original volume of the components. For mixtures of water and alcohol, the deficit volume is maximal for $V_{alcohol}/V_{water} = 3/2$. This could be a physical reason for the "exclusively mild" taste of good vodka. —A. A.

22.5   The role of wine in the prevention of cardio-vascular diseases: the French paradox (or Bordeaux effect)

We would like to precede this section by reminding the reader that:

⚠   **According to US legislation the use of alcoholic beverages is prohibited for persons under 21 years age.**

The necessity of such a rule is due to the negative effects of alcohol that are not mentioned here.

Figure 22.3 presents the mortality rate caused by cardio-vascular diseases (per 100,000 of the population) in different countries of the world versus the average animal fat consumption (in kilocalories per day). The correlation is obvious: the more animal fats consumed (cholesterol), the higher the mortality rate from cardio-vascular diseases (an almost linear increase). Nevertheless, there is a point that does not correlate. This point indicates the position of France: the French consume a noticeable amount of fats, but their mortality rate from cardio-vascular diseases is relatively low. Indeed, as we can observe, the French eat more fat than the British, although their mortality rate from heart attacks is almost four times lower.

This statistic was obtained from the MONICA project in 1992 and published in *The Lancet*. It was the CBS (Columbia Broadcasting System) anchor, however, who was the first to get behind these data. In 1991 he made them public under the catchy title of the "French paradox", which later also became known as the "Bordeaux effect". From the moment it was discovered, this anomaly was accredited to the regular consumption of noticeable amounts of red wine by the French, specifically in the Bordeaux Province. Further scientific research in the other red wine producing areas led to the unambiguous conclusion: "red wine consumption results in the noticeable lowering of the risk from cardio-vascular diseases". What is the reason behind this? This question was not easy to answer. The point is that red wine contains about 2,000 different substances: various acids, phenols, vanilla, and traces of almost all known minerals. Polyphenols, whose content in red wine is about 1 gram/liter, and Phytoalexin, which is found in grape skin, generated a special interest of the researchers. It was discovered that Phytoalexin contains trans-reserveratrol which, being a strong anti-oxidant, prevents brain cells from aging. Further research

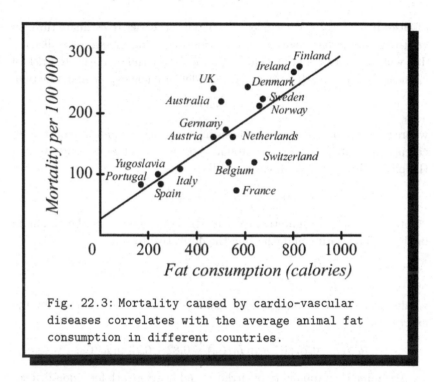

Fig. 22.3: Mortality caused by cardio-vascular diseases correlates with the average animal fat consumption in different countries.

showed that Polyphenols have an anti-lipoprotein impact, reduce the influence of Endotholin 1 (the major culprit of cardio-vascular diseases), and prevent the formation of platelets in the arteries. In this study we are not going any further into the jungle of medical science. Here we will try to estimate a healthy daily dose of red wine based on the physical methods of research. On the basis of the data from the MONICA report, we can assume that the probability of cardio-vascular diseases decreases with red wine consumption in accordance with an exponential law (as often occurs in nature):

$$I = I_0 \cdot e^{-b/b_i},$$

where I_0 is the probability for a non-drinker to contract the disease (e.g. for an Englishman, Figure 22.3) and b_i is the characteristic constant that can be estimated around a bottle a day.

It is obvious however, that one should not be overly enthusiastic about this method of heart attack prevention: too much alcohol can lead to serious diseases like cirrhosis of the liver. The danger is easier to estimate through vodka consumption: regular consumption of several glasses of vodka per day will have adverse health effects and will over time increase the risk of cirrhosis. Below we will again create a model based on exponential function:

$$C = C_0 e^{b/b_c},$$

where C_0 is the probability of cirrhosis in a non-drinker, and the constant b_c can approximately be assumed around three bottles of wine a day. Adding the probabilities of both diseases one has

$$W = I_0 \cdot e^{-b/b_i} + C_0 e^{b/b_c}.$$

"Optimum" wine consumption can be estimated to occur at the minimum in W. Differentiating and equating the derivative to zero

$$\frac{dW}{db} = -\frac{b}{b_i} I_0 \cdot e^{-b/b_i} + \frac{b}{b_c} C_0 e^{b/b_c} = 0,$$

we discover that the corresponding amount of wine is determined by the expression

$$b^*/b_i = 0.75 \left(1.1 + \ln I_0/C_0\right) = 0.82 + 0.75 \ln I_0/C_0.$$

Assuming the probability of cirrhosis and heart attack for a non-drinker being the same (the authors could not find the exact data) we can see that about half a liter of red wine a day is the "optimum" amount of consumption. This is the amount of wine that Tuscany farmers drink regularly, for instance[j].

22.6 Quality estimation and attribution of wine: SNIF-NMR method

When discussing the merits of responsible consumption of red wine endowed with certain properties, we must be sure that its origin matches what is

[j]Note that such generous daily consumption of wine is not a flight of imagination. In Heidelberg castle e.g., visitors can still find the world's largest barrel from which wine was pumped to the big dining hall by a special system of manually operated pumps. Average per person wine consumption in this castle, including children and old people amounted to 2 liters per day. The court jester, midget Perkey, would regularly drink 12 bottles a day. Cirrhosis of the liver was not the cause of his death. He died from dysentery after he drank a glass of untreated water, because he lost a bet.

written on the label. Contemporary physics offers the most accurate, NMR based methods to determine the origin of wine: SNIF (Specific Natural Isotope Fraction). Nuclear magnetic resonance (NMR) basically means the following: nuclei (for example, the protons in a hydrogen atom) have a small magnetic moment, a kind of microscopic compass. When placed in a homogeneous constant magnetic field **H** these magnetic moments precess around the direction of the field with a frequency ω_L proportional to the strength of H. Now let us consider what happens if one applies also a small radiofrequency (RF) field by means of a coil connected to a current generator and with an axis perpendicular to the field **H**. When the frequency of the RF field is just ω_L (now you understand why one speaks of "resonance"), electromagnetic energy is absorbed by the nuclei and consequently they reverse the direction of the magnetic moments with respect to the field **H**. This simple vectorial description is not far from the real phenomenon, although it is actually based on a complex quantum behavior of the elementary magnetic and angular moments of the nuclei (for more details see below in Chapter 28). The point we must emphasize here is that electronic equipment allows us to measure with very high precision the value of the magnetic field **H** at which the resonance occurs from the so-called "NMR resonance lines" of the absorption spectrum. The value of the resonance field is actually the local one at the nuclear site, namely the one resulting from the external field **H** and from small correction terms (very small indeed but still detectable) due to the electronic currents. Thus, it is clear that the resonance lines occur at different values of the electromagnetic irradiation frequency, depending on the environment of electrons around the nucleus. For instance, in ethyl alcohol, $CH_3 \, CH_2 \, OH$, one may expect resonance lines of intensities in the ratios 3 : 2 : 1 that occur at different frequencies in correspondence to the protons belonging to the molecular groups CH_3, or to CH_2 or to OH. Thus the resonance spectrum becomes a kind of snapshot of the molecular configuration. The SNIF method was devised by Gerard and Maryvonne Martin in Nantes in the 1980s, initially with the purpose of controlling the eventual enrichment of the wines by addition of sugar. In 1987, a company called Eurofins Scientific was founded and an *ad hoc* database was started. Today this database contains the NMR spectra of a large number of wines from France, Germany, Italy, and Spain. The SNIF method has been accepted by the European Community (1989) and by the Organization International de la Vigne e du Vin (OIV). It has also been acknowledged as the official method by the Association of Official

Analytical Chemistry (AOAC) in the USA and Canada and awarded "The Method of the Year" by the AOAC in 1996. Nowadays, the SNIF method makes it possible to single out ethyl alcohol with the same chemical structure but different botanic origin. It is possible to find out whether or not a wine originated from the must of a particular vine of a particular region. The method is based on the fact that, as a consequence of the various processes of photosynthesis, of the plant metabolism, and of the different geographical and climate conditions, the fraction of deuterium with respect to hydrogen is different from area to area and from vine to vine. The fraction of deuterium (D) with respect to hydrogen (H) is usually measured in part per million (ppm). While it is around 16,000 on Venus and 0.01 in the troposphere, on the earth the fraction D/H is around 90 at the South pole and 160 at the equatorial belt, and displays a relatively large variability. A further possibility for the characterization is related to the distribution of deuterium among different molecular groups. Again, for ethyl alcohol one can find $CH_2D - CH_2 - OH$ or $CH_3 - CDH - OH$ or $CH_3 - CH_2 - OD$. The percentage (D/H) for each molecular group can be estimated from the NMR spectra of deuterium, since the resonance lines occur at different frequencies because of the corrections to the external magnetic field due to the different electronic currents induced in each of the groups. From the NMR signals, compared by means of a suitable computational program with the spectra collected in the database, one can find out if sugar has been added to wine, determine the geographical origin of mixed wines versus the one claimed on the bottle label, etc.

PART 4

Windows to the quantum world

Finally the time has come to tell you about strange laws ruling in the world of microparticles. You will learn how these laws reveal themselves through "superphenomena" which take place at very low and not so low temperatures. We shall try to make things as easy as possible. Still this will not be small talk, for any real comprehension appeals to the mathematical language of modern physics. But we hope that you'll appreciate the marvels of this unbelievable, surprising, and promising world.

Chapter 23

The uncertainty principle

> Coordinate and momentum resemble the male and fe-
> male silhouettes in an antique barometer. For either of
> them to show up, the second must disappear.
>
> Werner Heisenberg.

In 1927 the German physicist Werner Heisenberg[a] discovered the uncertainty principle. Suppose that watching some body we managed to determine the projection of the momentum onto the x-axis with the accuracy Δp_x. Then we shall not be able to measure the corresponding x-coordinate with precision greater than $\Delta x \approx \hbar / \Delta p_x$, where $\hbar = 1.054 \cdot 10^{-34} \, J \cdot s$ is Planck's constant.

At first, this relation looks perplexing. Remember that according to school textbooks Newton's laws allow to write down equations of motion of a body and to calculate time-dependencies of the coordinates. Knowing those one can compute the velocity $\vec{v}$ (that is the time derivative of the coordinate $\vec{x}$), the momentum $\vec{p}$, and its projections. It looks as if we have established both coordinates and momenta and there is no uncertainty principle. Indeed, this is the fact in classical physics[b], but the situation changes in the microworld, Figure 23.1.

---

[a]W. K. Heisenberg, (1901–1976), one of the founders of quantum mechanics; 1932 Nobel Prize in physics.

[b]Of course the uncertainty relation holds in the macroworld too. But there it is not restrictive and does not play a key role.

Fig. 23.1: Fixing x-coordinate with accuracy equal to the width of the slit makes the corresponding projection of momentum P_x uncertain.

23.1 Momentum and coordinate

Imagine that we want to trace the motion of an electron. What should we do? The human eye can hardly do this. Its resolving ability is too weak to discern an electron. Then, let us try a microscope. The resolution of a microscope is limited by the wavelength of light used for the observation. Wavelengths of the usual visible light are of the order of $100\,nm$ $(10^{-7}\,m)$ and one can not see particles smaller than that in a microscope. Sizes of atoms are of the order of $10^{-10}\,m$ and there is no hope to discern them, not to mention a single electron.

But let us fantasize. Suppose that we managed to construct a microscope which exploits electromagnetic waves of smaller length, say, X-rays or even γ-rays instead of visible light. The harder the γ-radiation we use the shorter the corresponding waves and the smaller the objects which can be detected. This imaginary γ-microscope seems to be an ideal instrument capable of measuring electron positions with a desired accuracy. And what about the uncertainty principle?

Think over this hypothetical experiment (some physicists like the German word *Gedankenexperiment*) once more. In order to get the information about the position of the electron, at least one γ-quantum must be reflected. One quantum carries the minimal amount of energy of the radiation that is equal to $E = h\nu = \hbar\omega$ where ω is the angular frequency of the field oscillations. Short waves have higher frequencies and their quanta carry bigger energies. But the momentum of a quantum is proportional to the energy. Colliding with an electron a quantum inevitably transfers to the particle a fraction of its momentum. Because of that any coordinate measurement makes the electron momentum ambiguous. Rigorous analysis of the process proves that the product of the uncertainties can not become smaller than Planck's constant.

You may think that the discussion referred only to this particular case. Or we have proposed the wrong device and there are more delicate ways to measure the coordinate without kicking the electron to a new state. Unfortunately this is not so.

The best scientists (including Einstein) tried to invent a thought experiment which could determine the position and momentum of a body with accuracy better than what the uncertainty principle prescribed. But all attempts failed. By the laws of nature this is impossible[c].

Our arguments may appear vague and there is no self-evident mental model at hand. Real understanding demands a serious study of quantum mechanics. But hopefully the preceding was enough to make a first acquaintance with the subject.

In order to mark the boundary between the micro- and macroworlds let us make an estimate. Tiny particles used in observations of Brownian motion[d] are about $1\,\mu m$ ($10^{-6}\,m$)in size and weigh less than $10^{-10}\,g$. Still these fragments of matter contain enormous numbers of atoms. The uncertainty principle tells us that for them $\Delta v_x\,\Delta x \sim \hbar/m \sim 10^{-21}\,m^2/s$. Suppose that we are going to fix the particle position with an accuracy equal to 1 percent of the size, $\Delta x \sim 10^{-8}\,m$. Then $\Delta v_x \sim 10^{-13}\,m/s$. This is a very small quantity and the reason for that is the small value of Planck's constant.

The Brownian velocity of such a particle is approximately $10^{-6}\,m/s$. Apparently the inaccuracy of the velocity coming from the uncertainty relation is negligible. It is less than one tenmillionth (0.0000001!) even for a body this small. And because of the $\hbar/m$ in the right side of the relation it is even less apparent for larger bodies. But if we reduce the mass (take for example an electron) improving in the same time the accuracy of measurements (let $\Delta x \sim 10^{-10}\,m$, the atomic size) the uncertainty of the velocity becomes comparable to the velocity itself. For electrons in atoms

[c]Lately the attention of physicists was caught by so-called *squeezed states*. There the product of uncertainties is somewhat smaller but still of the order of $\hbar$. Existence of such exceptional states does not affect the general principle. —A. A.

[d]The chaotic motion of small impurities in gas or liquids due to collisions with their molecules. The average energy acquired by a dust particle because of these collisions is equal to the average thermal kinetic energy of a single molecule, $E = \frac{3}{2}kT$. —A. A.

the uncertainty principle carries full weight and may not be ignored. This leads to breath-taking consequences.

23.2 The probability waves

The simplest model of the atom is the Rutherford[e] planetary model where electrons circle the nucleus like the planets orbit the Sun. But electrons are charged particles and orbiting brings on varying electric and magnetic fields. This gives rise to electromagnetic radiation which dissipates energy. If the planetary model were correct, after having emitted all the energy, the electrons would fall onto the nucleus. While the planetary model predicts collapse of atoms, the stability of atoms is a solid experimental fact.

The Rutherford model required a "touch-up". This was done in 1913 by Niels Bohr[f]. In his model electrons are allowed to occupy only certain orbits with strictly defined energies. Electrons may change their energy only by jumping from one orbit to another. This "quantum" behavior explains many things and among those the atomic spectra and stability of atoms. Even now it is helpful in a simplified treatment of quantum effects. But it violates the uncertainty principle! Obviously, in contrast to the laws of the microworld both the coordinate and the momentum of an electron in orbit are definite, regardless of whether the orbit is quantum or classical.

A further development corrected this so-called semiclassical model. The actual behavior of electrons in atoms proved even more startling.

Suppose that we managed to find exactly where the electron is at a given moment[g]. Is it possible to predict positively where it will be a bit later, to be definite, say, in a second? No, because as we know, the coordinate measurement has inevitably introduced the uncertainty into the momentum. Predicting where the electron gets is beyond the power of devices. What should we do?

Let us mark the spatial point where we have found the electron. Another

---

[e]E. Rutherford, 1st Baron of Nelson, (1871–1937), English physicist; 1908 Nobel Prize in chemistry.

[f]Niels H. D. Bohr, (1885–1962), Danish physicist; the first to propose the idea of quantization; 1922 Nobel Prize in physics.

[g]Another problem we meet in quantum mechanics is that electrons are identical and it is impossible to distinguish them. The further discussion implicitly refers to the hydrogen atom which contains only one electron. —A. A.

mark will register the result of analogous measurement performed on one more atom of the type. The more measurements, the more marks. It turns out that although it is impossible to tell where the next one appears, the spatial distribution of the marks follows a pattern. The density of marks varies from point to point indicating the probability of meeting the electron.

We had to give up the idea of describing the motion of the electron in detail but we can judge the probability of locating it at different points in space. The behavior of an electron in the microworld is characterized by probability! The reader may dislike this strange suggestion. It is absolutely out of habit and contradicts our intuition and routine experience. But there is nothing we can do about the fundamentals of nature. The laws of the microworld are really different from those of every day. According to a comparison by Einstein we must "cast dice" in order to predict the behavior of electrons. One has to face the facts[h].

So, in the microworld the state of the electron is defined by the probability of finding it at various spatial points. In our pictorial model, the probability is proportional to the density of the marks. One may fancy that the marks form a sort of a cloud in which the electron lives.

But what regulates the structure of the probability clouds? You know that classical mechanics is ruled by Newton's laws. Quite similarly quantum mechanics has its own equation which determines the "spread" of an electron in space. This equation was found in 1925 by the Austrian physicist Erwin Schrödinger[i]. (Note that this was before the uncertainty principle clarified the reason for particle spreading. These things happen in physics.) The Schrödinger equation provides exact and detailed quantitative description of atomic effects. But it is impossible to solve it without complicated mathematics. Here we shall cite some ready answers that illustrate electron spreading.

Figure 23.2 presents a scheme of an experiment on electron diffraction. The pattern of bands which appears on the screen is shown in the photograph. It is indubitably akin to a pattern of light diffraction. The result would be impossible to explain if electrons followed linear trajectories as prescribed by the laws of classical physics. But if they are smeared in space this is conceivable. Moreover, the experiment demonstrates that probabil-

---

[h]We must note that Einstein himself did not believe in the necessity of such gambling. He did not accept quantum theory until the end of his life.

[i]E. Schrödinger, (1887–1961), Austrian physicist; 1933 Nobel Prize in physics.

ity clouds exhibit wave properties. Probability waves such as the birth rate or crime waves are common in day-to-day life. The amplitude of the wave is maximal in the place of the greatest probability of an event. In our case it is most probable to find the electron there. In the photograph these zones are lighter in color.

Fig. 23.2: Experiment on electron diffraction. 1 --- electron gun; 2 --- edge; 3 --- shadow; 4 --- fluorescent screen.

The smearing of the electron in a hydrogen atom as obtained by exact mathematical analysis of several quantum states is portrayed in Figure 23.3. These are analogues of quantum electron orbits in the Bohr model of the atom. Again the probability to meet the electron is higher in the lighter regions. The pictures are reminiscent of the snapshots of standing waves in finite domains. Probability clouds are really magnificent! Besides these abstract pictures do indeed define the behavior of electrons in an atom and explain, for example, energy levels and all that concerns chemical bonding.

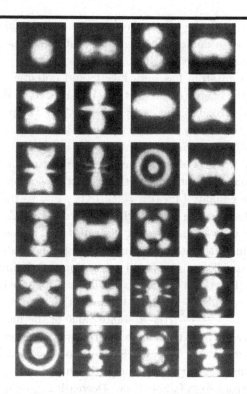

Fig. 23.3: The ''smearing'' of electrons in an atom. This is not a photograph but a result of calculations. Symmetries of these pictures strongly influence symmetries of molecules and crystals. One may even say that they are the keys to understanding of the beauty of ordered life forms in nature.

The uncertainty principle makes possible an estimate of the dimensions of probability clouds without going into the particulars of their structure. If the size of a cloud is of the order of Δx then it makes no sense to speak of coordinate uncertainties greater than Δx. Consequently, the uncertainty of momentum of the particle, Δp_x, can not be less than $\hbar / \Delta x$. The same expression determines by the order of magnitude the minimal momentum of the particle.

The smaller a cloud, the bigger the momentum and velocity of motion within the localization domain. It turns out that these general considerations suffice to make a correct estimate of the atomic size.

An electron in an atom has both kinetic and potential energies. The kinetic energy is the energy of the motion. It is related to the momentum by the well-known formula: $E_k = m v^2 / 2 = p^2 / 2m$. The potential energy of the electron is the energy of interaction with the nucleus. In the *International System of Units (SI)* it is equal to $E_p = -\frac{1}{4\pi\varepsilon_0} \frac{e^2}{r}$ where e is the charge of electron, r is the distance between the electron and the nucleus and the dimensional constant is called the permittivity of free space. It may be found from the equality $(4\pi\varepsilon_0)^{-1} = 9 \cdot 10^9 \, m \, / \, F = 9 \cdot 10^9 \, J \cdot m \, / \, C^2$.

The electron has definite value of the full energy $E = E_k + E_p$ in every state. The state with minimal energy is called the *ground (nonexcited)* state. Let us estimate the atomic radius for the ground state.

Imagine that the electron is spread over a domain of dimension r_0. Attraction to the nucleus tends to make r_0 smaller and collapse the probability cloud. This corresponds to lessening the potential energy that by an order of magnitude is $-\frac{1}{4\pi\varepsilon_0} \frac{e^2}{r_0}$ (the absolute value of the negative quantity grows as r_0 decreases). If that kinetic energy was not there the electron would fall onto the nucleus. However, as you remember, localized particles always possess kinetic energy by virtue of the uncertainty principle. And this prevents electrons from falling down! Decreasing r_0 enhances the minimal momentum of the particle $p_0 \sim \hbar / r_0$ and as a result the kinetic energy $E_k \sim \hbar^2 / 2m \, r_0^2$ grows too. The full energy of electron E is minimal when the derivative dE / dr_0 is zero. From this we obtain that the minimum corresponds to

$$r_0 \sim \frac{4\pi\varepsilon_0 \, \hbar^2}{m \, e^2}. \tag{23.1}$$

This determines the typical dimension of the localization domain that is essentially the atomic radius. The obtained value of r_0, (23.1) is $0.05 \, nm$ $(5 \cdot 10^{-11} \, m)$. As you know, by an order of magnitude this is the actual dimension of the atom. Obviously if the uncertainty principle makes it possible to correctly estimate atomic radius it must belong to the most profound laws of the microworld.

Another principle that follows directly from the uncertainty relation pertains to complex atoms. *Ionization energy E_i* is defined as the work required for detaching an electron from the atom. It can be measured pretty accurately. Imagine that the product of $\sqrt{E_i}$ and the atomic size d is the same for absolutely different atoms up to 10–20 percent. You may have already guessed the reason: the momentum of electron is $p \sim \sqrt{2mE}$ and according to the uncertainty relation the product $p \cdot d \sim \hbar$ must be constant.

23.3 Zero-point oscillations

Impressive results come out of applying the uncertainty principle to oscillations of atoms in the solid state. Atoms (or ions) oscillate about the nodes of a crystal lattice. Usually the oscillations are due to thermal motion and they increase as temperature rises. But what happens if the temperature is reduced? From a classical point of view the amplitude of oscillations will decrease and atoms will come to a halt at absolute zero. But is this possible from the point of view of quantum laws?

Shrinking the amplitude of oscillations means, in quantum language, compressing the probability cloud (or the localization domain) of a particle. We have seen, that because of the uncertainty relation, the price to pay will be the enhancement of the particle momentum. Attempts to arrest a quantum particle fail. It turns out that even at absolute zero temperature atoms in solids continue to oscillate. These *zero-point oscillations* give rise to a number of beautiful physical effects.

First of all let us try to estimate the energy of zero-point oscillations. In an oscillatory system a restoring force $F = -k\,x$ appears as the body is shifted to a small distance x away from the equilibrium. For a spring k is the elasticity coefficient while in a solid it is defined by forces of interactions between the atoms. The potential energy of the oscillator is

$$E_p = \frac{k\,x^2}{2} = \frac{m\,\omega^2\,x^2}{2},$$

where $\omega = \sqrt{k/m}$ is the frequency of oscillations.

This means that the energy of the oscillator may be expressed in terms of the amplitude of oscillations x_{max},

$$E = \frac{m\,\omega^2\,x_{\text{max}}^2}{2}; \qquad x_{\text{max}} = \sqrt{\frac{2E}{m\,\omega^2}}.$$

But in quantum language the amplitude of oscillations is the typical dimension of the localization domain that, because of the uncertainty relation, determines the minimal momentum of the particle. It comes out that, on the one hand, the smaller the energy of oscillations the smaller the amplitude must be. But, on the other hand, reducing the amplitude increases the momentum and, consequently, the kinetic energy of the particle. The minimal energy of a particle is given by the estimate,

$$E_0 \sim \frac{p_0^2}{2\,m} \sim \frac{\hbar^2}{m\,x_0^2} \sim \frac{\hbar^2}{m} \cdot \frac{m\,\omega^2}{E_0}.$$

Comparison of the last two expressions results in $E_0 \sim \hbar\omega$. The exact calculation gives half the value. The energy of zero-point oscillations is equal to $\hbar\omega/2$. It is maximal for light atoms that oscillate with higher frequencies.

Probably the brightest manifestation of zero-point oscillations is the existence of a liquid that does not freeze even at absolute zero. Obviously a liquid will not freeze if the kinetic energy of atomic oscillations is enough to destroy the lattice. It does not matter whether the kinetic energy appears due to the thermal motion or due to quantum oscillations. The most likely candidates for nonfreezing liquids are hydrogen and helium. The energy of zero-point oscillations is maximal in these lightest substances. Moreover, at normal temperature helium is an inert gas. The interaction between helium atoms is very weak and it is comparatively simple to melt the crystal lattice. It proves that the zero-point oscillation energy is enough and helium does not freeze even at absolute zero. On the contrary the interaction of atoms in hydrogen is much stronger and it freezes despite the zero-point oscillation energy of atoms being greater in hydrogen than in helium.

All other substances do freeze at absolute zero. So helium is the only one that always remains liquid at normal pressure. One may say that it is the uncertainty principle that prevents it from freezing. Physicists call helium a *quantum liquid*. Another exceptional property of helium is *superfluidity*, see page 244, which is also a macroscopic quantum phenomenon.

Still, under the pressure of about 2.5 *MPa* liquid helium becomes solid. However, solid helium is not an ordinary crystal. For example the kinetic energy of atoms at the interface of solid and liquid helium is defined by the zero-point oscillations. This enables the crystal surface to perform gigantic oscillations as if it was an interface of two nonmiscible fluids, Figure 23.4. Physicists have graced solid helium with the title of *quantum crystal* and vigorously investigate its properties.

Fig. 23.4: Waves on the interface between solid (the light region) and liquid helium.

Chapter 24

On snowballs, nuts, bubbles, and... liquid helium

Helium, despite its second position in the periodic table of elements, has, since its discovery, been a source of a great many hassles for physicists due to its quite unorthodox properties. Yet these troubles and headaches were overwhelmingly outshone by the beauty and uniqueness of the physical phenomena occurring in liquid helium as well as by the opportunities it offers to researchers and engineers in the production of extremely low temperatures. Besides superfluidity, among the peculiarities of this quantum liquid is its specific mechanism of charge transfer which differs from other liquids. And here our story starts.

Physicists started tinkering with this question in the late 1950s. At that time, the most probable candidates for the role of charge carriers seemed to be electrons and positive ions produced by the ionization of helium atoms. The further assumption was that electric charge is transported not by the helium ions themselves (they are rather heavy and it would be too hard to accelerate them) but by the *"holes"*. To get an idea of what the "holes" are, you may imagine an electron sitting in a helium atom has leaped onto a positively charged helium ion which happened nearby. When doing so, the electron must have left behind an empty place. However that "electron vacancy" may soon enough be occupied by another electron hopping from another atom. That new empty seat, in turn, will be taken by another electron continuing in this way. From aside, such an electron "leapfrog-game" looks as if a positively charged particle moved in the opposite direction. Yet, there is no real moving positive charge, just an absence of electron at its "dwelling place"; we can call this object a "hole". This charge transfer mechanism generally works fine in semiconductors, so it was considered quite plausible that it works in liquid helium.

As a rule, in order to measure the masses of charge carriers, both positive and negative, researchers study their trajectories in a uniform magnetic field. It's known that when a charged particle with some initial velocity enters a magnetic field it starts gyrating and its trajectory becomes a circle or a spiral. Knowing the initial velocity and the strength of the field one can easily find the mass of the particle simply by measuring the radius of gyration. However, the results of experiments turned out to be quite surprising: the masses of both the negative and positive charge carriers exceeded that of the free electron by tens of thousands of times!

Sure enough, in liquid, electrons and holes are surrounded and interact with atoms and, hence, their masses can differ from that of the free electron. Yet five orders of magnitude seemed way too much. Such a significant discrepancy of theoretical calculations and experimental results was considered unacceptable even for an extravagant element like helium. So there arose an urgent need to come up with a new model.

The American physicist Robert Atkinson soon proposed the correct explanation of the structure of the charge carriers in liquid helium. It is known that in order to transform a liquid into solid one doesn't necessarily need to cool it – it's just as possible to solidify liquid by compressing it. The pressure at which the liquid becomes solid is called the *solidification pressure* (P_s). Naturally, P_s depends on temperature: the higher the latter is the more difficult it gets to solidify the liquid by compression, and, therefore, P_s goes up. It turns out that the whole "trick" with the structure of the positive charge carriers is explained by the rather low value of solidification pressure of liquid helium: at low temperatures, $P_s = 25$ *atm*. And this causes the very unusual structure of the positive carriers.

We've mentioned earlier that positive ions, He^+, can commonly exist in liquid helium. When interacting with a neutral helium atom, a positive He^+ attracts the negatively charged electrons and repels the positively charged nucleus. The result is that the centers of positive and negative charges in the atom don't coincide anymore but get separated. Hence the presence of positive ions in liquid helium should lead to the *polarization* of its atoms. The polarized atoms are attracted by the positive ion and this, in turn, causes a rise of the local concentration of He atoms and the local density. As a result, the pressure around the ion increases. Graphically the dependence of the incremental pressure on distance from the positive ion is depicted in Figure 24.1.

Fig. 24.1: Polarized atoms are attracted by electric charge and the local pressure P of liquid helium is enhanced.

Well, as we already know, at low temperatures liquid helium solidifies at 25 atm. As soon as the pressure in the vicinity of the positive helium ion reaches that critical value, a certain volume around it turns solid[a]. According to Figure 24.1 for low external pressures this solidification takes place within approximately $r_0 = 0.7\,nm$ $(0.7 \cdot 10^{-9}\,m)$ from the ion. Therefore it gets "frozen" in a "snowball" formed by solid helium. Now, if an electric field is applied, the "snowball" will start moving. But it's not going to move alone and will be accompanied by its new retinue, pulling along a whole "tail" of extra density.

Consequently, the total mass of the positive charge carrier will include the three major contributions. The first one is the mass of the "snowball" itself, which is the product of the density of solid helium times the "snowball" volume at normal ambient pressure. This gives 32 m_0 ($m_0 = 6.7 \cdot 10^{-27}\,Kg$, is the mass of a helium atom). The mass of the following retinue turns out to be just slightly less — the mass of the tail of extra density pulled by the ion amounts to 28 m_0.

Besides these two, there must be one more mass added: when an object moves in liquid, there always occurs some displacement of masses of the liquid around it. This, of course, takes some energy. That is, to accelerate a body in liquid requires greater force than the same acceleration in a vacuum. So in liquid an object behaves as if its mass was somewhat greater.

[a]That, as you remember from Chapter 23, is rather flabby.

This additional mass due to the motion of liquid layers is called the *associated mass*[b]. For the "snowball" moving in the liquid helium at normal atmospheric pressure the associated mass turns out to be $15m_0$.

Finally, after summing up, the total mass of the positive charge moving in the liquid helium equals $75m_0$, a value which closely agrees with the experimentally measured one.

You see that the concepts of classical physics may deal successfully with the theory of positive charge carriers in liquid helium. But it is not so easy with negative ones. First of all, it happens that there are no negative ions in liquid helium at all (although a few of the negatively charged molecular ions He_2^- may be formed, they don't play any noticeable part in charge transfer). Hence, the electron is still the only remaining contestant for the role of the negative charge carrier. However, it catastrophically misses most of the mass required from the experimental data. And here is exactly the place for the idiosyncrasies of the quantum world to appear. Experiment shows that the electron, whom we have been so stubbornly intending for the negative charge carrier, can't even freely penetrate into liquid helium.

To make sense of all this, we will have to digress and touch a little on the structure of atoms with several electrons. There is a paramount principle, unquestionably reigning in the microworld, determining the behavior of groups of identical particles. When applied to electrons, it is called *Pauli's exclusion principle*[c]. According to this rule, no two electrons can occupy the same quantum state at the same time. And we shall show that this explains the observed "aversion" of helium atoms towards free electrons and the troubles the latter run into when they try entering the liquid helium.

The energy of an electron in an atom, as we've noted already, can have only certain quantum values. And what's important is that for each such a value of energy there are several corresponding states available for electrons, varying by the character of their motion in an atom (for instance, by the electron orbit shape or, in quantum language, the shape of the probability cloud which defines the spread of electrons in space, see Figure 23.3). States of the same energy compose a so-called *shell*. According to the Pauli principle, when the number of electrons in an atom grows (as the atomic

---

[b]We have encountered this concept already in Chapter 20 (see page 169).

[c]Wolfgang Pauli, (1900–1958), Austrian born, lived in Germany, Switzerland, and in the US; specialized in quantum mechanics, quantum field theory, relativity, and other fields of theoretical physics; 1945 Nobel Prize in physics.

number goes up), they do not get "crammed" in the same states but fill new available shells one by one.

The first shell, corresponding to the lowest possible energy, must be occupied first. Located real close to the atomic nucleus, it can take in only two electrons. Thus, in helium, which is the second in the periodic table, the first shell is completely filled. There is no choice for the third electron other than to stay sufficiently far from the nucleus. When such an "undue" electron approaches a helium atom within a distance of the order of its radius, there appears a repulsion force precluding it from getting near.

Therefore, some "entrance work" is needed to ram an errant electron into the bulk of helium. Three Italian physicists, Carreri, Fasoli, and Gaeta, proposed the idea that as the electron entering helium can't get too close to the atoms, it shoves them away and thereby forms a spherically symmetrical cavity, some sort of a bubble, Figure 24.2. And this bubble with the electron scurrying inside is indeed the negative charge carrier in liquid helium.

Fig. 24.2: Because of quantum effects electrons can not approach He atoms and instead shove them away.

The size of the bubble can be rather easily estimated. The repulsion between electrons and helium atoms should decrease with distance. On the other hand, at large distances, electrons should act on helium atoms in exactly the same way the positive ions do, that is to polarize them. So, the interaction of electrons with helium atoms should be the same attraction as for the previous situation with "snowballs". Hence, when approaching a bubble with the electron captured inside, the incremental pressure in helium rises, following the law portrayed in Figure 24.1.

Yet, under normal conditions the pressure at the boundary of the bubble still remains far less than $25\,atm$ because of the comparatively large size of the bubble. Besides this pressure, resulting from the increased density of polarized helium, there is a force of surface tension. It acts at the bubble's border and is directed in the same way, to the center of the bubble. What would then balance these two external forces preventing our bubbles from collapse? It turns out that the required counteraction is created by that very same imprisoned electron.

Indeed, according to the uncertainty principle, which we have discussed in Chapter 23, the accuracy of measuring the momentum of the electron is directly related to the uncertainty of the electron's position in space, being $\Delta p \sim \hbar / \Delta x$. In our case the uncertainty in location of the electron is naturally defined by the size of the bubble, that is $\Delta x \sim 2R$. So, the movement of the captive electron should posses a momentum of the order of $\hbar / 2R$ and, consequently, have a kinetic energy $E_k = p^2/2m_e \sim \hbar^2 / 8m_e R^2$. Resulting from the electron's collisions with the walls of the bubble, there should arise some outward pressure (remember the principal equation of the kinetic theory of gases, relating the pressure of gas P to the average kinetic energy of its chaotically moving particles and their density: $P = \frac{2}{3}nE_k$). This pressure could very well balance the forces trying to squeeze our helium bubble. In other words, electron confined in the bubble acts in exactly the same way as gas isolated in a reservoir; however, this "electron gas" consists of a single particle! The density of such a gas is, obviously, $n = 1 / V = 3 / 4\pi R^3$. After plugging that value and $E_k \approx \hbar^2 / 8m_e R^2$ into the expression for the pressure, we find that $P_e \approx \hbar^2 /16\pi m_e R^5$. Precise quantum mechanical calculations lead to a similar answer[d]: $P_e \approx \pi^2 \hbar^2 /4m_e R^5$.

As long as the external pressure remains small, the dominating force trying to squeeze the bubble will be surface tension, $P_L = 2\sigma / R$, see Chapter 10. Hence, equating $P_e = P_L$, one can easily estimate the radius of a stable electron bubble in liquid helium:

$$R_0 = \left(\frac{\pi^2 \hbar^2}{8 m_e \sigma} \right)^{\frac{1}{4}} \approx 2\,nm = 2 \cdot 10^{-9}\,m.$$

We see now that the negative electrical charge in liquid helium is carried

---

[d]The origin of the discrepancy is that the electron prefers to stay in the middle of the cavity rather than near the repelling walls. This effectively reduces the uncertainty of coordinate and enhances the pressure.

by bubbles with electrons ensnared inside.

The total mass of such carriers can be calculated in the same manner as was carried out for the snowballs. Yet, now the bubble itself weighs almost nothing for the mass of the electron inside is negligibly small compared to the mass of the liquid dragged by the bubble (the retinue) plus the associated mass. So, the net mass of the carrier would be equal to the sum of the associated mass and the mass of the tail pulled by the drifting bubble. Because of the rather large size of the bubble, its resulting mass, $245\,m_0$, turns out to be much greater than that of the snowball.

Now, let us consider how an increase in external pressure P_0, will effect the properties of the charge carriers. Figure 24.1 depicts the dependence of the total pressure (including the external one), $P = P(r) + P_0$, in the vicinity of an ion in liquid helium versus distance from the ion for $P_0 = 20\,atm$. Such dependence for an arbitrary value of $P_0 < 25\,atm$ can be graphed by simply shifting the curve for $P_0 = 0$, along the P-axis. As the picture shows, the higher the external pressure is, the further from the ion the total pressure attains $25\,atm$. Hence, with the growth of the external pressure, the snowball behaves as if it was rolling down a snowy slope: it promptly lumps up on itself the "snow" – solid helium – and gets bigger. The relation between the size of the snowball and the value of external pressure $r(P_0)$, is presented in Figure 24.3.

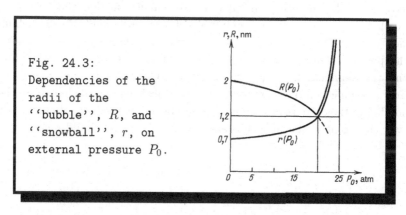

Fig. 24.3:
Dependencies of the radii of the ''bubble'', R, and ''snowball'', r, on external pressure P_0.

And what about the "bubble"? How does it behave as P_0 goes up? Well, for some while, like any bubble in liquid, it submissively shrinks while the surrounding pressure grows. Its radius $R(P_0)$ decreases as the upper curve in Figure 24.3 shows. And yet, further, at $P_0 = 20\,atm$ the curves for $r(P_0)$

and $R(P_0)$ cross, meaning that at this point the sizes of the bubble and the snowball become the same and equal 1.2 *nm*. We know already the future fate of the snowball: with the rise of P_0 it will rapidly grow at the expense of the helium solidifying on its surface. But what is the bubble to do — continue to shrink following the dashed line in Figure 24.3?

No way! At this very moment, our bubble finally shows its real character. As the applied pressure continues to go up, the bubble begins to act as a snowball: it gets covered with an icy crust of the solid helium. Indeed, according to Figure 24.1 at $P_0 = 20$ *atm* the total pressure on the surface of the bubble becomes equal to 25 *atm*, reaching the solidification point for liquid helium. The internal radius of the bubble, protected now by its fancy icy attire, ceases to change and remains approximately the same regardless of the further increase of pressure, whereas its external radius equals that of the snowball at the corresponding pressure.

Thus, at external pressures greater than $P_0 = 20$ *atm*, our bubbles become coated with an ice shell and begin to resemble nuts. With that difference, though, that the kernel in the "nut" is of a quite peculiar nature – it's an electron rushing chaotically in its shell of solid helium.

One last thing worth mentioning here. As P_0 approaches 25 *atm*, the external radii of both the nuts and the snowballs continue to grow bigger (trying, in principle, to reach infinity). Finally all the helium in the reservoir becomes solid. The role of negative charge carriers in the solid helium is, therefore, played by the electron bubbles, frozen in the bulk of solid helium and having inherited their dimension of about 1.2 *nm* from the former nuts. The positive charge, on the other hand, must be transferred by helium ions, the remnants of the former snowballs. Of course, it's not so easy to carry anything in a rigid environment, the electric charges included. So the mobility of the carriers in solid helium will be by many orders of magnitude lower than that of the snowballs and the bubbles in the liquid phase.

Chapter 25

Superconductivity passion at the end of the Millennium

Probably almost all our readers have heard of superconductivity. This phenomenon consists of the abrupt disappearance of electrical resistance of some pure metals and alloys at low temperatures. "Low temperatures" meaning the range of 10–$20\,K$, that is 10–20 degrees above absolute zero ($-273.15°\,C$). In order to cool to these low temperatures a sample is usually placed into liquid helium that at normal pressure boils at $4.2\,K$ and, as you already know, does not freeze down to absolute zero. Throughout the last century physicists and chemists in many laboratories all over the world were looking for compounds which become superconducting at high enough temperatures to be cooled, for instance, by comparatively cheap and widely available liquid nitrogen. So understand that the discovery of high-temperature superconductors, whose resistance becomes zero at temperatures above $100\,K$, was met as the greatest event in physics of recent years. Really, the practical significance of this discovery can be compared to that of magnetic induction at the beginning of the nineteenth century. It ranks with the discovery of uranium fission, the invention of the laser, and the discovery of the unusual properties of semiconductors in the twentieth century.

25.1 Starting from the end

The beginning of this exciting new stage in the development of superconductivity was the work by K. A. Müller and T. J. Bednorz at IBM's lab in Switzerland. In the winter of 1985–86 they managed to synthesize a compound of barium, lanthanum, copper, and oxygen – the so-called

metal oxide ceramic La − Ba − Cu − O, a compound which had superconducting properties at $35\,K$, the record at that time. The article, cau-
tiously titled "The Possibility of High-Temperature Superconductivity in
the La − Ba − Cu − O System", was turned down by the leading American
journal *Physical Review Letters.* In the previous 20 years the scientific community had gotten tired of receiving sensational reports about the discovery
of high-temperature superconductors that turned out to be false, so it decided to avoid the trouble. Müller and Bednorz sent the article to the German journal *Zeitschrift fur Physik.* When the news about high-temperature
superconductivity was finally heard and research began in hundreds of laboratories, every article devoted to the new phenomenon would start from
a reference to this article. But in the fall of 1986 it passed practically un-
noticed. Just one Japanese group checked the result and verified it. Soon
the phenomenon of high-temperature superconductivity was corroborated
by physicists in the United States, China, and the Soviet Union.

At the beginning of 1987 the whole world was in a fever, searching for
new superconductors and investigating the properties of those already discovered. The critical temperature T_c increased quickly: it was $T = 45\,K$ for
La − Sr − Cu − O and it reached $52\,K$ for La − Ba − Cu − O under pressure. Finally in February 1987 the American physicist Paul Chu got the
idea to imitate the effect of external pressure by substituting La atoms by
the smaller atoms of Y which is next in Mendeleev's table column. The
critical temperature of the compound Y − Ba − Cu − O (see Figure 25.1)
broke the "nitrogen barrier", reaching $93\,K$. This was a long-awaited triumph but not the end of the story. In 1988 a five-component compound of
the type Ba − Ca − Sr − Cu − O with the critical temperature $110\,K$ was
synthesized and a little later its mercury and thallium analogues with the
critical temperature $125\,K$ appeared. The maximum critical temperature,
under pressure of $30\,GPa$, of the mercury record-breaker looks impressive
even on the Celsius scale $-108°\,C$!

The discovery of high-temperature superconductivity is unique in modern physics. First, it was discovered by just two scientists with very modest
tools. Secondly, the compounds include easily accessible elements. As a
matter of fact, these superconductors can be made in a high school chemistry lab in a day. What a contrast with discoveries in other areas such
as high-energy particle physics which require massive collaborations and
millions of dollars worth of equipment. There, a list of authors alone takes
a whole page in a journal article. This discovery was a cause for optimism:

Fig. 25.1: Structure
of superconducting
$Y Ba_2 Cu_3 O_7$.

Ba   Y   Cu   O

the time of lone investigators in physics hasn't passed! And though the discovery had been anticipated for 75 years, it caught everyone by surprise. Theorists could only shrug their shoulders and continue to do so as the critical temperature went up.

So was the discovery by Bednorz and Müller a fluke or destiny? Could the discovered compound, with its unique properties, have been synthesized earlier? How difficult is it to answer these questions! We have long been accustomed to the fact that everything new is obtained on the edge of the impossible by using unique equipment, superstrong fields, ultralow temperatures, superhigh energies... There is nothing of the kind here. It isn't too difficult to "bake" a high-temperature superconductor – a qualified alchemist of the Middle Ages could have managed it. It's worth recalling that about 10 years ago many laboratories of the world intensively investigated an unusual superconducting compound. This substance was called "alchemic gold" because of its yellow luster and high density, which made it resemble the noble metal. It was synthesized by medieval alchemists, passed off as true gold, and advertised as the result of successfully using the "philosopher's stone". Alchemic gold is a complex compound, and who knows, perhaps a high-temperature superconductor would have been baked in the Middle Ages if it had been blessed with a golden luster.

The dreams of the Middle Ages may take us too far. But you will be amused to hear that some of the present-day high-temperature superconductors were in fact stored on lab shelves since 1979! At that time they were synthesized in quite a different connection at the Moscow Institute of General and Inorganic Chemistry by I. S. Shaplygin and his collaborators. Unfortunately they did not measure the conductivity of the compositions at low temperature which would indicate the new phenomenon and the discovery did not happen[a].

25.2   From surprise to understanding

Now that all the world speaks of the properties and prospects of high-temperature superconductors many steps in the history of superconductivity appear in a new light.

Superconductivity, one of the most interesting and unusual phenomena in solid-state physics, first became known on April 28, 1911, at a meeting of the Royal Academy of Sciences in Amsterdam, when the Dutch physicist Heike Kamerlingh Onnes[b] reported a recently discovered effect: the complete disappearance of electrical resistance of mercury cooled by liquid helium to $4.15\,K$. Though no one expected this discovery, and it contradicted the existing classical electron theory of metals, the fact that it was Kamerlingh Onnes who discovered superconductivity was not accidental. Actually, he was the first scientist who managed to solve the most complicated scientific and technical problem of the time: obtaining liquid helium (which boils at $4.16\,K$). This allowed scientists to peek into the unknown world of temperatures close to absolute zero. Kamerlingh Onnes immediately tried to apply the new experimental means and to investigate the

[a]A superconductor baked in the Middle Ages would have been doomed to oblivion for the same reason: there was no liquid helium at hand! —A. A.

[b]H. Kamerlingh Onnes, (1853–1926), Dutch physicist; Nobel Prize 1913 in physics.

low-temperature behavior of pure metals. This was the time of hot theoretical debate over whether resistance of pure metals turns to zero or remains finite at absolute zero. Being an advocate of the first theory Kamerlingh Onnes was clearly satisfied by the result that he had obtained for mercury. But soon he realized that the vanishing of resistance at finite temperature is an effect quite different from the expected one.

We'd like to emphasize that the resistance of a sample in the superconducting state is equal to zero not approximately but exactly. That's why electric current in a closed circuit can circulate as long as you like without damping. The maximal duration of a persistent superconducting current recorded in England was about two years. (The current in the ring would have circulated up till now but for a transport worker strike of which caused a break in the supply of liquid helium to the laboratory.) Even after two years, no damping of the current was detected.

Very soon superconductivity was discovered not only in mercury but in other metals as well. The prospects for practical applications of the discovered phenomenon seemed unlimited: power transmission lines without waste, superpowerful magnets, electric motors, new types of transformers, but there were two obstacles to such inventions.

The first were the extremely low temperatures at which superconductivity was observed in all materials known at the time. To cool conductors to these temperatures, scarce helium is used (its stocks are limited, and even now producing a liter of liquid helium costs some dollars). This makes many projects using superconductivity simply unprofitable. The second obstacle discovered by Kamerlingh Onnes was that superconductivity had turned out to be rather sensitive to magnetic fields and to the value of current. In fact, it was destroyed by strong fields.

The next fundamental property of the superconducting state discovered in 1933 was the Meißner-Ochsenfeld effect: the complete expulsion of magnetic field from the volume of the superconductor. But again experimental investigations were complicated by the need to work with scarce liquid helium – before World War II it was produced in about 10 laboratories throughout the world (two of which were in the Soviet Union).

The fundamentals of superconductivity stayed absolutely out of reach of the classical theory of metals and the quantum one was still an embryo at that time. The so-called two-liquid model suggested a coexistence of two types of electrons in superconducting metals: normal electrons interact with the lattice but superconducting ones for some reason don't. This assumption let the London brothers[c] write down the equations of electrodynamics of superconductors that described the Meißner effect and some other features. Still the microscopic mechanism of superconductivity remained a mystery.

In 1938 P. L. Kapitza[d] discovered the phenomenon of superfluidity. It turned out that at temperatures below $2.18\,K$ liquid helium can flow through thin capillary tubes without any viscosity. The theoretical explanation of this phenomenon by L. D. Landau[e] gave rise to hopes that a theory of superconductivity was in the offing. It turned out that helium atoms at low temperature behave like quantum particles with whole spin and get accumulated at the lowest energy level (the *Bose-condensation*). Landau showed that a gap appears as the result in the spectrum of excitations making possible the superfluid state. Discussing this macroscopic revelation of this entirely quantum effect Landau called helium "a window to the quantum world".

A straightforward extension of these ideas to superconductivity failed. The reason was that electrons are particles with one-half spin (so-called *fermions*) and behave absolutely unlike helium atoms which possess a whole spin being *bosons*. In the quantum system of electrons, excitations with zero energy may appear even at zero temperature and the Landau criterion of superfluidity does not hold.

The natural desire to reduce the problem to that already solved inspired the idea to prepare from two fermions a composite boson with a whole total spin and after that to effect the Landau superfluidity scenario. However this was opposed by the Coulomb[f] repulsion of electrons which was too strong in spite of the screening that occurs in an electroneutral metal.

Ten years later, in 1950, the discovery of the *"isotopic effect"* first indicated the connection between superconductivity and the crystal lattice of a metal. Measurements of the critical temperature of lead proved that it depended on the mass number of the isotope under testing. Thus superconductivity ceased being a purely electronic phenomenon. A little later

---

[c]H. London, (1907–1970), British physicist; F. London, (1900–1954), American physicist; specialists in low-temperature physics.

[d]P. L. Kapitza, (1894–1984), Russian physicist; 1978 Nobel Prize in physics.

[e]L. D. Landau, (1908–1968), Russian physicist; 1962 Nobel Prize in physics.

[f]C. A. de Coulomb, (1736–1806), French physicist and inventor; discovered the basic law of electrostatics in 1785.

Frölich[g] and Bardeen[h] independently demonstrated that the interaction of electrons with lattice oscillations *(phonons)* may lead to an attraction. This could in principle overcome the electrostatic repulsion but one has to keep in mind the huge kinetic energies of electrons. At first glance those should overcome the just-mentioned weak coupling. Composite bosons did not work out.

Later in 1950, with the help of experimental data and the theoretical achievements of solid-state physics, based on quantum mechanics and statistical physics, Ginzburg[i] and Landau (USSR) developed a phenomenological theory of superconductivity, known as the Ginzburg–Landau theory. It proved so successful and predictive that even now it remains a powerful research tool despite the 50 elapsed years since its creation.

In 1957 the American scientists John Bardeen, Leon Cooper, and Robert Schrieffer[j] put together these ideas and hints and created a consistent microscopic theory of superconductivity. It was found that superconductivity is indeed linked with the appearance of a peculiar attraction of electrons in metals. This is an utterly quantum phenomenon.

We have already mentioned that the ground state of a fermionic system is characterized by big kinetic energies of electrons. Luckily those do not prevent the binding of low-energy excitations of the system that behave like *quasiparticles*. They have the same electric charge e as an electron and some effective mass but their energy may be whatever small. The attraction brings on a rearrangement of the quasiparticle spectrum and the long-awaited gap that was so crucial for the Landau superfluidity criterion opens at last.

The origin of the attraction may be understood with the help of an analogy of two balls lying on a rubber rug. If the balls are far from each other, each of them deforms the rug, making a little depression. But if we put a ball on the rug and place another one near the first, their holes will join, the balls will roll down to the bottom of the combined valley and

[g]H. Frölich, (1905–1991), born in Germany, worked in Leningrad, 1933–1935 and since 1935 in Britain.

[h]J. Bardeen, (1908–1991), American physicist; 1956, 1972 Nobel Prize in physics(!)

[i]V. L. Ginzburg, (1916–2009), Russian physicist and astrophysicist; 2003 Nobel Prize in physics.

[j]L. Cooper (born 1930), and R. Schrieffer, (born 1931), American physicists; 1972 Nobel Prize in physics (together with J. Bardeen).

lie together. In metals, the mechanism is realized by deformations of the crystal lattice. At low temperatures some quasiparticles (usually they are called, just the same, electrons) form bound pairs. These are called "Cooper pairs" after the man who discovered the binding. The size of the pairs on the atomic scale is really quite large, reaching hundreds and thousands of interatomic distances. According to the graphic comparison suggested by Schrieffer, they should be envisaged not as a double star composed of electrons but rather like a couple of friends in a discotheque who either come together or dance in different corners of the hall, separated by dozens of other dancers.

You see that it took almost half a century since the discovery of superconductivity to gain cardinal progress in understanding its nature and to develop a consistent theory. This period may be considered to be the first stage of superconductivity studies.

25.3   Chasing high critical parameters

The creation of the theory of superconductivity was a powerful impulse to investigate it in earnest. Without fear of overstatement, we can say that great progress has been achieved in producing new superconducting materials in the subsequent years. The Soviet scientist A. A. Abrikosov's[k] discovery of an unusual superconducting state in a magnetic field played a significant role in this development. Before then a magnetic field was thought to be incapable of penetrating the superconducting phase without destroying it (which is actually true for most pure metals)[l]. Abrikosov theoretically proved that there was another possibility: under certain conditions a magnetic field could penetrate into a superconductor in the form of current vortices. The core of the vortex turned to the normal state but the periphery remained superconducting! Depending on their behavior in a magnetic field, superconductors were divided into two groups: superconductors of the first type (old) and those of the second type (discovered by Abrikosov). It's important that superconductors of the first type can be

---

[k]A. A. Abrikosov, (born 1928), Russian physicist, pupil of L. D. Landau, specialist in condensed matter physics; 2003 Nobel Prize in physics.

[l]Strictly speaking this is true only for cylindrical specimens placed in a magnetic field parallel to the axis of the cylinder. If either the specimen is not a cylinder or a strong enough field is oriented differently the so-called intermediate state may be realized. It is formed by alternating macroscopic layers of superconducting and normal phases.

changed into one of the second type if we "spoil" them by adding impurities or other defects.

A real hunt for superconducting materials with high critical fields and temperatures started. The ingenuity of the pursuers was really boundless. Arc welding, instant cooling, and sputtering onto hot substrate were employed. The efforts resulted in the discoveries of, for example, the alloys Nb_3 Se and Nb_3 Al which have the critical temperature (that is the temperature of transition to the superconducting state) $T_c = 18\,K$ and the upper critical field of more than $20\,T$. Further progress was achieved lately with ternary compounds. Before the discovery of high-temperature superconductors the record value of the upper critical field $(60\,T)$ belonged to the alloy $Pb\,Mo_6\,S_8$ with $T_c = 15\,K$.

Among superconductors of the second type, scientists managed to find compounds capable of carrying a high-density current and bearing gigantic magnetic fields. And although many problems had to be solved before they could find practical application (the compounds were brittle, high currents were unstable), the fact remained: one of the two major obstacles to the widespread use of superconductors in technology was overcome.

But increasing the critical temperature was still problematic. If critical magnetic fields were increased thousands of times in comparison with Kamerlingh Onnes's first experiments, the changes in critical temperature weren't too encouraging: at best it managed to reach $20\,K$. So for the normal operation of superconducting instruments the expensive liquid helium was still necessary. This was particularly vexing because a fundamentally new quantum effect, the "Josephson effect", had been discovered. This made it possible to use superconductors widely in microelectronics, medicine, instrumentation, and computers.

The problem of increasing the critical temperature was extremely acute. Theoretical evaluations of its peak value showed that in the context of normal phonon superconductivity (that is, superconductivity caused by electron attraction due to interaction with the crystal lattice), this temperature could not exceed $40\,K$. But the discovery of a superconductor with such a critical temperature would still be a great success, since that could be achieved with relatively cheap and available liquid hydrogen (which boils at $20\,K$). It would open the era of "mid-temperature superconductivity". This stimulated attempts to modify existing superconductors and create new ones by traditional methods of material science. But the ultimate dream was to create a superconductor with a critical temperature of $100\,K$

(or, even better, above room temperature), which could be cooled by cheap and widely used liquid nitrogen.

The best result of the search was the alloy $Nb_3 Ge$ with the critical temperature of $23.2\,K$. This record temperature was achieved in 1973 and stood for 13 years. Until 1986 the critical temperature couldn't be raised by even one degree. It seemed that the possibilities of the phonon mechanism of superconductivity had been exhausted. In view of this, in 1964 the American physicist Little and the Soviet scientist V. L. Ginzburg proposed the following idea: if the possibility of increasing the critical temperature is limited by the nature of the phonon mechanism of superconductivity then it should be replaced by something else – that is, electrons should form Cooper pairs by means of some other, nonphonon, attraction.

During the last 20 years many theories were proposed, and tens or hundreds of thousands of new substances were investigated in detail. In his work Little has drawn attention to quasi-one-dimensional compounds – long molecular conducting chains with side branches. According to theoretical evaluations, a noticeable increase in critical temperature could be expected there. Despite the attempts of many laboratories throughout the world, such superconductors were not synthesized. But on the way physicists and chemists made many wonderful discoveries: they obtained organic metals, and in 1980 crystals of organic superconductors were synthesized (the current record for the critical temperature of an organic superconductor is over $10\,K$). They managed to obtain two-dimensional layered metal-semiconductor "sandwiches" and at last magnetic superconductors where the former enemies, superconductivity and magnetism, coexisted peacefully. But there were no new prospects for high-temperature superconductivity.

By this time superconductors had extended the range of application, but the need to cool them with liquid helium remained the weak spot.

In the mid-1970s strange ceramic compounds of the type $Pb - Ba - O$ appeared as candidates for high-temperature superconductivity. In their electrical properties they were "poor metals" at room temperature but became superconducting not too far from absolute zero. "Not too far" means about 10 degrees below the record value of the time. But the new compound could hardly be called a metal. According to theory, the obtained value of critical temperature was not by any means low, but surprisingly high for such substances.

This attracted attention to ceramics as would be high-temperature superconductors. Since 1983 Müller and Bednorz worked like alchemists with hundreds of different oxides, varying their composition, quantity, and conditions of synthesis. According to Müller they were led by some physical ideas that are now being validated by the most complicated experimental studies of the new materials. In this painstaking way they stealthily approached the compound of barium, lanthanum, copper, and oxygen that showed superconducting properties at 35 K.

25.4 Quasi-two-dimensional superconductivity between antiferromagnetic and metallic states

A fair number of various chemical compounds exhibiting temperatures of superconducting transition higher than the record of 1973 have been synthesized by now. The chemical formulae of some that have critical temperatures above the boiling point for liquid nitrogen are summarized in Figure 25.2.

The feature shared by high-temperature superconductors is the layered structure. The best studied high-temperature superconductor is by now the compound $Y Ba_2 Cu_3 O_7$. Its crystal structure is illustrated in Figure 25.1 on page 241. It is easily seen that atoms of copper and oxygen are arranged in planes interspaced by others atoms. As a result the conducting layers are separated by insulating ones and the motion of charge carriers (those are usually not electrons but holes) is *quasi-two-dimensional*. Namely, holes migrate freely within $Cu O_2$ layers although hops between the layers are comparatively rare. Cooper pairs are also localized in the planes.

Apparently the quasi-two-dimensional nature of the electron spectrum in high-temperature superconductors is a key to the understanding of the microscopic mechanism of this wonderful phenomenon. This has yet to be realized. Nevertheless a brilliant phenomenological theory of the vortex state in high-temperature superconductors is already at hand. It proved so rich with diverse effects that now, in fact, it constitutes a new realm of physics, i.e. the physics of "vortex matter". A bedrock of that is the quasi-two-dimensionality of the electron liquid.

Indeed, once electrons and Cooper pairs are confined in two dimensions Abrikosov vortices consist of elementary vortices which are attached to the conducting planes. These elementary vortices are known among physicists as "pancakes". At low temperature the pancakes draw up in a line due to

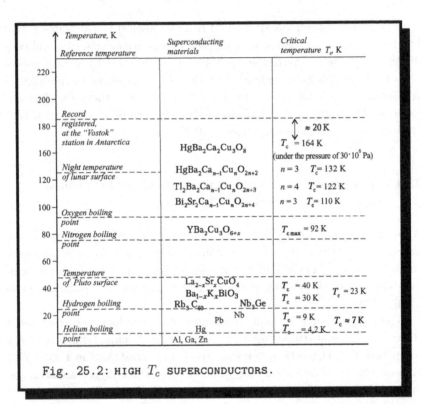

Fig. 25.2: HIGH T_c SUPERCONDUCTORS.

a weak attraction between them. Then the lines form a vortex lattice. As temperature rises thermal fluctuations make the vortex lines more twisted and at some point the vortex lattice melts almost as if it was an ordinary crystal. Thus in a high-temperature superconductor the ordered Abrikosov lattice gives way to a disordered "vortex liquid" phase formed by chaotically twisting tangled vortex lines. It is interesting that the further growth of temperature may break apart vortex lines and cause vortices to "evaporate" but at the same time preserve the superconductivity. Elementary vortices in the layers will become absolutely independent of each other and of vortex configurations in neighboring planes. Inhomogeneities of various kinds that are inevitably present in real crystals make the phase picture of the vortex matter even more complicated.

Despite significant progress in understanding of properties of high-temperature superconductors the mechanism behind the effect remains a secret. No less than 20 conflicting theories proclaim having explained high-temperature superconductivity but what we need is the one that is true.

Some physicists believe that Cooper pairs in these superconductors are formed because of a magnetic fluctuation interaction of some sort. An indication may be the fact that the critical temperature and the concentration of free electrons drop down in crystals, $Y\,Ba_2\,Cu_3\,O_{6+x}$, impoverished in oxygen, that is at $x < 1$, Figure 25.3, (the right curve). At $x < 0.4$ one deals already with a dielectric, but at low enough temperatures a magnetic ordering of copper atoms takes place. The magnetic moments of neighboring atoms become anti-parallel and the total magnetization of the crystal stays zero. This type of ordering is well known in the physics of magnetism where it is called *antiferromagnetic ordering* (see the left curve in Figure 25.3; here T_N is the so-called Néel[m] temperature, that is the temperature of the transition to the antiferromagnetic state). One could believe that copper atoms retain the fluctuating magnetic moment in the superconducting phase and in the long run this gives origin to the superconducting attraction of electrons. This mechanism leans on specific properties of copper atoms which, depending on the valence, are either magnetic or not. The presence of $Cu - O$ layers in all high-temperature superconductors could be considered as an argument in favor of the theory. However quite recently the superconductivity of $W_3\,O\,Na_{0.05}$ at $90\,K$ was reported. The exact composition of the superconducting phase is not known yet; or for certain there are no "magic" copper atoms in it. Moreover, none of the elements in the formula of the new high-temperature superconductor show magnetic properties.

In other theories physicists try to generalize in one way or another the classical theory of superconductivity, revise the very basics of theory of the metallic state, "crossbreed" superconductivity and ferromagnetism in spaces of higher dimension, separate spin and charge of carriers, preformed Cooper pairs in advance at temperatures higher than critical, and undertake other attempts to explain the unusual properties of the high-temperature superconductors in a universal manner.

The challenge of nature waits for the answer and theoretical community can not come to accord. On the one hand, this may confirm the famous comparison with a chorus of the deaf where everyone performs his own part with disregard of the others. But on the other, it may be the case that the time has not yet come and the correct theory is not yet formulated.

---

[m]L. Néel, (1904–2000), French physicist; 1970 Nobel Prize in physics.

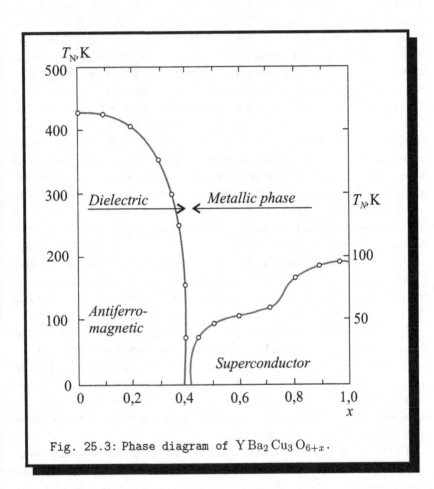

Fig. 25.3: Phase diagram of $YBa_2Cu_3O_{6+x}$.

Chapter 26

What is a SQUID?

26.1 The quantization of magnetic flux

In the microworld, that is the world of atoms, molecules, and elementary particles, many physical quantities may take only definite discrete values. Physicists say that they are quantized. (We have already mentioned that according to the Bohr rule, energies of electrons in atoms are quantized[a].) Macroscopic bodies consist of big collectives of particles and chaotic thermal motion leads to an averaging of physical quantities. This smears the little steps and conceals quantum effects at a macroscopic level.

Now, what if the body is cooled to a very low temperature? Then arrays of microparticles can move in accord and reveal quantization at macroscopic scales. A good example is the fascinating phenomenon of quantization of magnetic flux.

Everyone who has studied the laws of electromagnetic induction knows the definition of the magnetic flux through a closed contour:

$$\Phi = B\,S,$$

where B is the value of magnetic induction and S is the area encircled by the contour (for simplicity let the field be normal to its plane). Nevertheless it will be a discovery for many that magnetic flux produced by superconducting current, say, in a ring may assume only discrete values. Let us try to understand this at least superficially. Presently it's sufficient to believe that microparticles are moving along in quantum orbits. This simplified

[a]See page 222.

image often substitutes probability clouds in blackboard discussions.

The motion of superconducting electrons in a ring, Figure 26.1, resembles that of electrons in atoms: it seems that electrons follow gigantic orbits of the radius R without any collisions. Therefore a natural assumption is that their motion obeys the same rule as in atoms. The Bohr postulate states that only certain orbits of electrons are stationary and stable. They are selected by the following quantization rule: products of the momentum of an electron mv and the radius of the orbit R (this quantity is called the angular momentum of the electron) form a discrete sequence:

$$mv\,R = n\,\hbar. \tag{26.1}$$

Here n is a natural number and $\hbar$ is the minimal increment *(quantum)* of the angular momentum, equal to the Planck constant $\hbar$. We have already met it when talking about the uncertainty principle. It turns out that the quantization of all physical quantities is determined by this universal constant.

Fig. 26.1: Electron in conducting ring.

Let us find the value of the magnetic flux quantum. Consider a single electron and let the magnetic flux through the ring gradually increase. As you know the electromotive force of induction appears:

$$\mathcal{E}_i = -\frac{\Delta\Phi}{\Delta t},$$

and the strength of the electric field is

$$E = \frac{\mathcal{E}_i}{2\pi R} = -\frac{\Delta\Phi}{2\pi R\,\Delta t}.$$

By Newton's second law the acceleration of a charged particle is:

$$ma = m\frac{\Delta v}{\Delta t} = -\frac{e\,\Delta\Phi}{2\pi R\,\Delta t},$$

where e is the electric charge. After the obvious cancelation of Δt we obtain:

$$\Delta\Phi = -\frac{2\pi m\,\Delta v\,R}{e} = -\frac{2\pi}{e}\Delta\,(m\,v\,R).$$

You see that magnetic flux across the ring is proportional to the angular momentum of electrons[b]. According to the Bohr quantization rule (26.1) angular momentum may take only discreet values. This means that magnetic flux through a ring with superconducting current must be quantized as well:

$$m\,v\,R = n\,\hbar \qquad \text{and} \qquad \Phi = (-)\frac{2\pi}{e}\,n\,\hbar. \tag{26.2}$$

The value of the quantum is extremely small ($\sim 10^{-15}\,Wb$) but twentieth-century technique makes it possible to observe magnetic flux quantization. The studies carried out in 1961 by the Americans B. S. Deaver and W. M. Fairbank used a superconducting current circulated not in a ring but in a hollow superconducting cylinder[c]. The experiment confirmed that the magnetic flux through the cylinder changed stepwise but the measured value of the quantum was half of that obtained above. The modern theory of superconductivity gives the answer. Remember that in the superconducting state electrons join into Cooper pairs with the charge $2e$. Superconducting current is a motion of these pairs. Therefore the correct value of the magnetic flux quantum Φ_0 is obtained when substituting into the formula (26.2) the electric charge $2e$ of a pair:

$$\Phi_0 = \frac{2\pi\hbar}{2e} = 2.07 \times 10^{-15}\,Wb.$$

This is the way to recover the factor two. We were not the first to miss it. The English theoretician F. London had lost it as well. He predicted the magnetic flux quantization as early as 1950, long before the nature of the superconducting state was understood.

[b]The relation between encircled magnetic flux and angular momentum of electrons is valid in both classical and quantum physics. —A. A.

[c]The quantization of magnetic flux was simultaneously observed by R. Doll and M. Näbauer. The two papers appeared in the same issue of *Physical Review Letters*.

It is worth saying that our derivation of the magnetic flux quantization certainly is too naive. It is rather surprising that we contrived to obtain the right meaning of the quantum this way[d]. In fact superconductivity is a complicated quantum effect. Those who really want to comprehend it have a long and hard way ahead of them. It demands many years of resolute but rewarding work.

26.2   Josephson effect

Let us turn to another quantum superconducting phenomenon that lay a cornerstone for several unrivaled measuring methods. The Josephson effect was discovered in 1962 by a 22-year-old British graduate student and brought him the Nobel Prize for the theoretical prediction 11 years later[e].

Imagine a glass plate (that is called a substrate) supporting a superconducting film. (Usually the superconducting material is sputtered in vacuum.) The surface of the film has been oxidized and the oxide forms a thin dielectric layer on it. Finally the superconductor is sputtered once again. The final outcome is a so-called superconducting sandwich interlaid by a thin insulating sheet. Sandwiches are widely used in observations of the Josephson effect. For convenience the two thin superconducting strips usually cross each other, see Figure 26.2.

Fig. 26.2: Josephson junction: 1 --- metal film; 2 --- oxide layer; 3 --- substrate.

We shall begin with a case when the metallic layers are in the normal, nonsuperconducting state. Is it possible for electrons to pass from one

---

[d]The weak points of our derivation are: first, it is impossible to change the magnetic flux in a superconducting ring due to the "freezing" (see later); second, superconducting pairs form a quantum collective state and there is no way to pick out a single pair. — A. A.

[e]B. D. Josephson, (born 1940), British physicist; 1973 Nobel Prize in physics.

metallic film into another, Figure 26.3, *a*?

Fig. 26.3: Potential
energy of electrons in
tunnel junction
without voltage.

From the first sight, no, because of the dielectric in between. The dependence of the electron energy versus the x-coordinate (X-axis is perpendicular to the plane of the sandwich) is plotted in Figure 26.3, *b*. Electrons in metal move freely and their potential energy is zero. The potential energy of electrons in dielectric, W_u, surpasses their kinetic (and total) energy in metal W_e. The work to be done by electrons when exiting to the dielectric[f] is $W_u - W_e > 0$. Therefore one says that electrons in the two films are separated by the potential barrier of the height $W_u - W_e$.

If the electrons obeyed the laws of classical mechanics the barrier would be insuperable. But electrons are microparticles and the specific laws of the microworld permit many things that would be ruled out for bigger bodies. For example neither man nor electron are able to mount a barrier higher than their energy. But an electron may simply penetrate through it! As if it tunneled under a mountain when the energy was not enough to climb it. This is called the tunnel effect or quantum tunneling. Of course you should not take this literally like really digging a hole. The true explanation comes from the wave properties of microparticles and their "spreading" in space, a deep understanding of which requires good command of quantum mechanics. But in a word electrons can with some probability pass through

---

[f]This resembles the heat of evaporation, that is the work done when extracting a molecule from liquid.

the dielectric from one metal film to another. The probability increases for
smaller heights $W_u - W_e$ and widths a of the barrier.

Once the dielectric film is permeable for electrons we may ponder the
electric current flowing through it. At the moment this so-called tunnel
current is zero: the number of electrons coming to the upper electrode
from below is equal to those going back.

What should we do to make the tunnel current nonzero? Simply break
the symmetry. For example, let us connect the metal films to a battery
with the voltage U, Figure 26.4, a. Then the films will act like two plates
of a capacitor and the electric field of the strength $E = U/a$ will set up
in the dielectric layer. The work done when moving a charge e a distance
x along the field is $A = Fx = eEx = eU\,x/a$ and the potential energy
of electrons takes the form plotted in Figure 26.4, b. Evidently electrons
from the upper film ($x > a$) penetrate the barrier more easily because those
moving from below must jump to the higher level. Therefore even small
voltages break the balance and give rise to a tunnel current.

Fig. 26.4: Potential
energy of electrons in
tunnel junction with
voltage.

Tunnel junctions of normal metals are used in electronic devices but
don't forget that our aim was the practical applications of superconductivity. The next step is to assume that the metal strips separated by the
insulating layer are superconducting. How does the superconducting tunnel junction behave? It turns out that superconductivity leads to quite
unexpected results.

We said that electrons in the upper film possess the surplus energy eU

with respect to the lower one. Upon coming down they must dump the energy and come to equilibrium with the others. This was not a problem in the normal state: several collisions with the crystal lattice would redistribute the extra energy and convert it to heat. But if the film is superconducting this way is not acceptable! It must emit the energy in the form of a quantum of electromagnetic radiation. The energy of the quantum is proportional to the applied voltage U:

$$\hbar \omega = 2e\,U.$$

You see that the electric charge in the right-hand side is twice that of the electron. This indicates that tunneling of superconducting pairs has taken place.

This was the dazzling prediction by Josephson. Applying constant voltage to a superconducting tunnel junction (sometimes called the Josephson junction) brings about the generation of electromagnetic radiation. The first experimental observation of this effect was performed in 1965 by I. M. Dmitrienko, V. M. Svistunov, and I. K. Yansons at the Kharkov Physical-Technical Institute of Low Temperatures.

The first use that comes to mind for the Josephson effect is the generation of electromagnetic waves. However it is rather difficult to extract the radiation from the narrow space between the superconducting films (this was a serious obstacle to experimental observation of the effect). Besides the emission is too weak. Now Josephson elements are used mainly as detectors of electromagnetic radiation because they are the most sensitive ones in certain frequency ranges.

This application exploits the resonance between the frequency of the external (registered) wave and the proper frequency of oscillations in the junction under a voltage. The idea of resonance is basic for most receivers: a set is "tuned in" when the proper frequency of the receiving contour is adjusted to that of the station. A Josephson junction makes a convenient receiving cell. The two advantages are: first, the frequency depends on the voltage and is easily varied; second, the resonance being very sharp results in high selectivity and precision. Josephson elements are employed in the most sensitive detectors for the observations of the electromagnetic radiation of the Universe.

26.3   The quantum magnetometer

The Josephson effect together with magnetic flux quantization provide a basis for a whole family of supersensitive measuring devices called *SQUIDs*. This abbreviation stands for *Superconducting Quantum Interference Devices*. We will now look at the quantum magnetometer, which measures weak magnetic fields.

The simplest quantum magnetometer consists of a superconducting ring with a Josephson junction, Figure 26.5. As you know, in order to create a current through a normal tunnel contact one must apply some voltage. But for a superconducting junction this is not necessary. Superconducting pairs may tunnel through the insulating layer and the superconducting current may circulate in the ring regardless of the Josephson junction. This is called the stationary Josephson effect. (In distinction to the nonstationary Josephson effect accompanied by emission that was described in the previous section.) However the current is limited by a maximum allowable value called the critical current of the junction, I_c. Currents exceeding I_c destroy the superconductivity of the junction and a voltage drop appears across it. The Josephson effect becomes nonstationary.

Fig. 26.5: Electric current and magnetic flux through superconducting ring with weak link.

So, the insertion of a Josephson junction does not completely destroy superconductivity of the contour. Nevertheless a segment of imperfect superconductivity, the so-called weak link, appears. It plays a crucial role in the operation of a quantum magnetometer. Let us try to understand this.

If the entire contour was superconducting the magnetic flux through it, Φ_{int}, would be strictly constant. Indeed, by the law of electromagnetic induction any change of external magnetic field gives rise to the electromotive force of induction, $\mathcal{E}_i = -\Delta\Phi_{\text{ext}}/\Delta t$, that effects the electric current. The change of the current in its own turn generates the electromotive force of selfinduction, $\mathcal{E}_{si} = -L\,\Delta I/\Delta t$. The resistance of the superconducting contour and the voltage drop in it are zero:

$$\mathcal{E}_i + \mathcal{E}_{si} = 0,$$

and

$$\frac{\Delta\Phi_{\text{ext}}}{\Delta t} + L\frac{\Delta I}{\Delta t} = 0.$$

Remember that the magnetic flux through the contour that arises due to the current I is $\Phi_I = LI$. This means that $\Delta\Phi_{\text{int}} = \Delta\Phi_{\text{ext}} + \Delta\Phi_I = 0$ and the change of the superconducting current compensates for the change of the external field. The total magnetic flux through the contour remains constant, $\Phi_{\text{int}} = \Phi_{\text{ext}} + \Phi_I = const$. There is no way to change it without transferring the contour to the normal state. The magnetic flux is "frozen".

What happens if the contour contains a weak link? Then magnetic flux through a contour may change since the weak link allows magnetic quanta to penetrate inside the ring. (You remember that magnetic flux encircled by a superconducting current is quantized and is equal to a whole number of the quanta Φ_0.)

Let us watch the magnetic flux through a superconducting ring with a weak link and the electric current in it as the external magnetic field changes. Let the initial external field and current be zero, Figure 26.5, *a*. Then the magnetic flux through the ring is zero as well. If we enhance the external magnetic field a superconducting current will arise and the external flux will be completely compensated. This will happen until the electric current reaches the critical value I_c, Figure 26.5, *b*. To be definite we shall assume that this happens when the flux of the external field equals one half of the quantum: $\Phi_0/2$.[g]

[g]Critical current depends on many factors and in particular on the thickness of the dielectric. It is always possible to fit the latter so that the flux created by the critical current has the desired value: $L\,I_c = \Phi_0/2$. This simplifies analysis but does not affect essentials.

As soon as the value of the current reaches I_c the superconductivity of the weak link is destroyed and the quantum of magnetic flux Φ_0 enters the ring, Figure 26.5, *c*. The ratio Φ_{int}/Φ_0 stepwise increases by unity. (The superconducting contour passes to the next quantum state.) And what about the current? The value remains the same but the direction reverses. Judge for yourself, formerly the external flux was compensated by the field of the current: $\Phi_I + \Phi_{\text{ext}} = -L I_c + \Phi_0/2 = 0$. After the quantum has entered the ring the current and the external flux add up: $\Phi_{\text{ext}} + \Phi_I' = \Phi_0/2 + L I_c' = \Phi_0$. Thus allowing the flux quantum in the system instantaneously has changed the direction of the current.

As the external field grows the current in the ring decreases and the superconductivity of the junction is restored. When the external flux is Φ_0 the current disappears, Figure 26.5, *d*. After that it changes direction again in order to screen the excess of magnetic flux. Finally, when the external flux comes to $3\Phi_0/2$ the current becomes I_c, the superconductivity of the junction is destroyed, and one more quantum of magnetic flux enters the ring, etc.

The dependencies of the magnetic flux across the ring, Φ_{int}, and the electric current I versus the external magnetic flux Φ_{ext} are shown in Figure 26.6. Both fluxes are measured in magnetic quanta Φ_0 which present the natural units. The stepwise shape of the dependence offers the possibility of "counting" individual flux quanta despite their extremely small ($\sim 10^{-15}$ *Wb*) magnitude. The reason is quite clear. Even though the magnetic flux through the superconducting contour changes by a tiny amount $\Delta\Phi = \Phi_0$ this happens in a very short time, Δt, almost instantly. Therefore the velocity $\Delta\Phi/\Delta t$ during this abrupt change may be really big. It can be measured, for example, by electromotive force induced in a special measuring coil of the device. This is the principle of operation of a quantum magnetometer.

Construction of a real quantum magnetometer is much more complicated. Say, usually not one but several weak links are connected in parallel. This gives rise to interference of superconducting currents (or, to be exact, the interference of the quantum waves that determine locations of electrons). This helps to increase precision of measurements. The collective name *SQUID* of such devices refers to the interference of quantum waves. The sensitive element of the device is inductively coupled with an oscillatory contour where the jumps in magnetic flow are converted into electric

Fig. 26.6: Magnetic flux *(a)* through superconducting ring with weak link and electric current *(b)* in it as a function of external magnetic flux.

impulses to be amplified later. But these technical subtleties are far beyond the scope of the book.

The fact is that supersensitive magnetometers capable of measuring magnetic fields with 10^{-15} T accuracy are now widely applied in industrial production. Among other things they are used in medicine. It turns out that the working heart, brain, and muscles create weak magnetic fields. For example the magnetic induction due to activity of the heart is $B \approx 10^{-11}$ T being a hundred thousand times less than the field of the Earth. But still these fields, however weak they are, lie within the reach of SQUIDs. Records of rhythms of these fields are called magnetocardiograms, magnetoencephalograms, etc. Superconducting facilities offer new possibilities to register and study the most delicate signals of the human organism. This was a breakthrough in medical diagnostics for many diseases.

Experiments in the field started in the 1970s. In order to minimize the influence of the magnetic field of the Earth measurements were carried out in specially designed screened chambers. Their walls were made of three layers of metal with high magnetic permeability that presented efficient magnetic screening plus two layers of aluminum in between for electric screening. These precautions provided the means to reduce magnetic field inside the chamber to several *nanoTesla* ($1\,nT = 10^{-9}\,T$) that is ten thousand times less than that of the Earth. Clearly such chambers cost a

fortune. Further development of this promising realm of SQUID application
led to remarkable progress and greatly simplified the procedure. Modern
superconducting technique permits taking a distinct magnetocardiogram
with no screens at all, Figure 26.7. The only imperative condition is to
remove metal clips and the content of your shirt pocket.

Fig. 26.7: Modern magnetocardiogram.

Even the "triple protection" did not eliminate
traces of the Earth's magnetic field in magnetocar-
diographic chambers. What could have been other
objectives for building them?

Chapter 27

Superconducting magnets

Strong magnetic fields can be obtained by passing strong electric currents through a coil. The greater the current the bigger is the field. If the coil possesses electric resistance, heat is released as the current flows. Supporting the current requires enormous energy and, besides, one must dissipate the heat which otherwise may fuse the coil. In 1937 a magnetic field with the induction $10\,T$ was first realized. The field could be supported only at night when few consumers were using power. The liberated heat was removed by running water and 5 liters (1.3 *gal*) of it were brought to the boil every second. The heat release set the main limitation to creating strong magnetic fields with ordinary coils.

As soon as superconductivity was discovered the idea appeared to exploit it in the production of strong magnetic fields. At first glance the only thing to be done is to wind up a coil of superconducting wire, send a sufficiently strong current, and short the circuit. Once the resistance of the coil is zero no heat is released. The gains would justify the work done when cooling the solenoid down to the temperature of liquid helium. But unfortunately, the strong magnetic field destroyed the superconductivity of superconductors that were known at that time.

A way out was found. The help came from laws of quantum mechanics. As you know, in superconductivity these laws may work on macroscopic scales.

27.1 The Meißner effect in detail

In Figure 27.1 you can see the scheme of the experiment that was performed
by Kamerlingh Onnes in 1911 in Leiden. The Dutch scientist put a lead
coil into liquid helium where it cooled down to helium's boiling point. The
electric resistance of the coil disappeared because it achieved the superconducting state. After that he reconnected the switch and closed the coil onto
itself. The undamped current began to circulate in the coil.

Fig. 27.1: Electric
current can circulate
in a superconducting
coil for years without
damping.

The current generates magnetic field with the induction proportional to
its strength. A naive assumption is that the larger the current in the coil
the bigger the magnetic field it produces. But the results were discouraging:
as the field reached several hundredths of a *Tesla* the solenoid passed to
the normal state and electric resistance appeared. Attempts were made to
prepare coils of other superconductors but in those again superconductivity
was destroyed at relatively weak fields. What was the rub?

The puzzle of this inconvenient behavior of superconductors was solved
in 1933 in the laboratory of W. Meißner in Berlin. It was found that superconductors possess the property of expelling magnetic field; the induction
inside superconductors is zero. Imagine that a metal cylinder (a piece of
wire) was cooled and became superconducting. Then one switched on a
magnetic field with the induction $\vec{B}_{ext}$. By the law of electromagnetic induction this must cause circular currents at the surface of the cylinder,
Figure 27.2. The magnetic field $\vec{B}_{cur}$ created by the currents inside the
cylinder is equal to $\vec{B}_{ext}$ in magnitude but opposite in direction. The cur-

rents are superconducting and do not die out. Therefore the net induction in the superconductor is zero: $\vec{B} = \vec{B}_{\text{ext}} + \vec{B}_{\text{cur}} = 0$. Lines of magnetic induction do not penetrate superconductors.

Fig. 27.2: Surface currents keep magnetic field out of superconductor of the first type.

But what if we change the order and apply the field before cooling the specimen to the superconducting state? It seems that the magnetic induction will not change and there would be no point in generating surface currents. This was the logic of Meißner when he checked calculations by Laue[a] concerning the first experimental procedure. But still he preferred to check. The result of the renewed experiment was stunning. It turned out that the magnetic field was just the same, forced out of the superconductor without penetrating it. This was called the Meißner effect.

Now it is clear why magnetic field destroys superconductivity. Exciting surface currents takes energy. In this sense the superconducting state is less favorable than the normal one when the magnetic field enters the bulk and there are no surface currents. The higher the induction of external field the stronger screening current it demands. At some value of magnetic induction the superconductivity will inevitably be destroyed and the metal will return to normal. The value of the field when the destruction of superconductivity occurs is called the critical field of the superconductor. It is important that the presence of an external field is not a necessary condition of the destruction. Electric current in the superconductor produces a magnetic field of its own. At a certain intensity of current the induction of the field

---

[a]M. von. Laue, (1879–1960), German physicist; 1914 Nobel Prize in physics.

reaches a critical value and the superconductivity breaks down. The value of the critical field increases at low temperatures but even near absolute zero critical fields of pure superconductors are modest, see Figure 27.3. So it could seem a vain hope to obtain strong magnetic fields with the help of superconductors.

Fig. 27.3: Critical magnetic field grows at low temperatures.

But further investigations in the field proved that the situation is not desperate. It was found that there is a whole group of materials that stay superconducting even in very strong magnetic fields.

27.2 The Abrikosov vortices

As was already mentioned above, in 1957 the Russian theoretical physicist A. A. Abrikosov (see page 246) showed that magnetic field does not destroy superconductivity of alloys so easily. Similar to the pure case, the magnetic field begins penetrating the superconductor at some critical value of induction. But in alloys the field does not occupy the entire volume of the superconductor at once. At first only detached bundles of magnetic lines are formed in the bulk, Figure 27.4. Every bundle carries an exactly fixed portion. It is equal to the quantum of magnetic flux, $\Phi_0 = 2 \cdot 10^{-15}$ Wb, that we have already met[b].

---

[b]It would be quite natural to say that each magnetic quantum corresponds to one line of magnetic induction. —*A. A.*

Fig. 27.4: Bundles of magnetic lines in superconductor of the second type.

The stronger the magnetic field the more bundles enter the superconductor. Each of them brings one magnetic quantum and the total flux changes stepwise. Again, as before, magnetic flux through a superconductor may take only discrete values. It is astonishing to see the laws of quantum mechanics working on macroscopic scales.

Each bundle of magnetic lines piercing the superconductor is enveloped by undamped circular currents that resemble a vortex in gas or liquid, Figure 27.4. For this reason the bundles of magnetic lines together with the superconducting currents around it are called Abrikosov vortices. Certainly in the core of the vortex, the superconductivity is broken. But in the space between the vortices it is conserved! Only in very strong fields when numerous vortices begin overlapping is the superconductivity destroyed completely.

This remarkable reaction of superconducting alloys to magnetic fields was first discovered at the tip of a pen. But modern experimental technique makes it possible to observe Abrikosov vortices directly. Fine magnetic powder is applied to the surface of a superconductor (for example, to the base of a cylinder). The particles gather at the places where the field enters the alloy. Electron microscope study of the surface reveals the dark spots.

Such a photograph of the structure of Abrikosov vortices is shown in Figure 27.5. We notice that the vortices are arranged periodically and form a pattern similar to a crystal lattice. The vortex lattice is triangular (this means that it is can be made up of periodically repeated triangles).

So, in distinction to pure metals, these alloys possess not one but two critical fields: the lower critical field marks the moment when the first

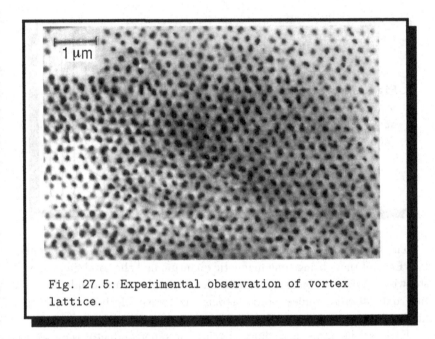

Fig. 27.5: Experimental observation of vortex lattice.

vortex enters the superconductor and the upper critical field corresponds to the complete destruction of superconductivity. Over the interval between these two the superconductor is pierced by vortex lines. This is called the *mixed state*. Superconductors exhibiting such properties are now called "type two". The first type refers to those where the magnetic destruction of superconductivity happens immediately.

It looked like the problem of producing superconducting magnets was solved. But nature had kept for researchers one more catch. The wire for superconducting solenoids must withstand not only strong magnetic field but strong electric current as well. As we shall see in the next section this happened to be different.

27.3 What is pinning?

It is well known that a force acts on an electric current in a magnetic field. But where is the counteraction force that must appear according to Newton's third law? When the field is caused by another current then, no doubt, it experiences an equal force in the opposite direction (the interaction of energized conductors obeys Ampère's[c] law). But our case is more difficult.

A current that flows in a mixed state superconductor interacts with the magnetic field in the cores of vortices. This affects the distribution of current but the domains where the magnetic field concentrate do not remain intact either. They start moving. Electric current compels Abrikosov vortices to move!

The force exerted onto a current by magnetic field is perpendicular to the magnetic induction and to the conductor. The force acting on Abrikosov vortices is also perpendicular to the induction of the field and to the direction of the current. Suppose that a current traverses the superconductor depicted in Figure 27.5 from left to right. Then the Abrikosov vortices will move either up- or downwards depending on the direction of the magnetic field. However, the transport of the Abrikosov vortex across the superconductor is the motion of the normal non-superconducting core. It undergoes friction which results in heat and dissipation. Current in the mixed state superconductor meets a resistance just the same. It could look like these materials are no good for solenoids.

What was the solution? It is to block the motion and hold the vortices in place. Fortunately this is possible. One has simply to build defects into the superconductor and worsen it. Usually defects appear by themselves as a result of mechanical or thermal treatment. Figure 27.6 demonstrates an electron-microscope photograph of niobium nitride film. The critical temperature of the film is $15\,K$. It was obtained by means of sputtering metal onto glass plate. One clearly discerns the granular (or rather columnar) structure of the material. It is not so easy for a vortex to jump the boundary of a grain. Hence up to a certain current strength, the so-called critical current, vortices stay in place and the electrical resistance is zero.

This phenomenon is known as *pinning*, because the vortices are pinned by defects.

Pinning offers the possibility of preparing superconducting materials exhibiting high critical values of both magnetic field and electric current. (It would be more accurate to speak not of critical current but of critical current density, that is the current crossing a unit area of cross section.) Critical field is determined by properties of material. In the mean time critical current depends on methods used in preparation and treatment of

---

[c]A. M. Ampère, (1775–1836), French physicist, one of the founders of classical electrodynamics.

Fig. 27.6: Microscopic structure of niobium nitride film.

the conductor. Modern technology provides a means to obtain superconductors with high values of all critical parameters. For example, starting from the tin-niobium alloy one can fabricate a material with the density of critical current reaching several hundreds A/cm^2, the upper critical field equal to $25\,T$ and the critical temperature being $18\,K$.

But this is not the end of the story. It is important that the mechanical properties of the material allow it to make a coil. The tin-niobium alloy by itself is too fragile and it would be impossible to bend such a wire. So the following procedure was invented: a copper tube was stuffed with a mixture of niobium and tin powders, then the tube was stretched into a wire and the coil was wound and heated making the powders fuse. This resulted into a solenoid of the Nb_3Sn alloy.

Industry prefers more practical materials such as the more plastic niobium-titanium alloy $NbTi$. It is used as a base for so-called composite superconductors.

First one drills a number of parallel channels in a copper bar and inserts the superconducting rods. Then the bar is stretched into a long wire. The wire is cut and the pieces inserted into another drilled copper bar. That is once more stretched, cut into pieces, and so on... Finally one obtains a cable that contains up to a million superconducting lines, like those shown in Figure 27.7. This is used for winding coils.

Fig. 27.7: Industrial cables are made of copper with millions of superconducting lines in it.

The important advantage of such cables is that the superconducting current is distributed among all the lines. When compared to a superconductor copper behaves like an insulator. If copper and a superconductor are connected in parallel then the entire current will choose the path that has no resistance. There is a second advantage. Suppose that by accident superconductivity breaks down in one of the lines. This causes heat liberation and the danger that the whole cable will pass to a normal state. It is imperative to remove the heat. Copper is a good heat conductor and perfectly suits the purpose of thermal stabilization as well as having good mechanical properties for cable.

27.4 *Postscriptum* for taxpayers

After having started with the high-temperature thriller we turned to applications of conventional superconductors. Here, in contrast to high-temperature superconductors, the physics of the phenomenon is clear. Nevertheless the lack of theoretical understanding does not stop the search for practical applications of high-temperature superconductors. The main stumbling block are poor technological properties of the available high-temperature superconductors: they are extremely brittle and do not stand rolling which is an essential element of mechanical treatment of metals. Nevertheless several brands of high-temperature superconducting cables that are kilometers long are already on the market. They are produced by rolling and annealing of a tube of silver or other suitable metal filled with high-temperature superconductor powder. A number of experimental underground transmission lines made of such cables are in operation now in France and the USA. The first electric motors and generators based on high-temperature superconductors are under testing. There is no doubt that the field of applications of these materials will expand and new, more practical high-temperature superconductors will appear.

Let us turn to the prospects. These are really fantastic. Many global projects of the past are put back on the agenda because of the advent of high-temperature superconductivity. For example, at present 20–30 percent of all produced electrical energy is wasted in power transmission lines. Using high-temperature superconductors in energy transmission could eliminate these losses.

All projects involving thermonuclear synthesis need giant superconducting magnets that keep high-temperature plasma away from the walls of the chamber. Streams, if not rivers, of liquid helium are necessary to maintain the superconducting state. The helium would be replaced by nitrogen at a tremendous cost saving.

Gigantic superconducting coils would serve as accumulators of electrical power, which would share the load during peak periods.

Supersensitive equipment for making magnetocardiograms and magnetoencephalograms, based on the use of superconducting Josephson elements, would come to every hospital.

Magnetic cushions created by superconducting coils would support intercity express trains commuting at speeds of $400-500\, km/h$.

A new generation of supercomputers based on superconducting elements and cooled by liquid nitrogen would be constructed.

Don't think we've lost our heads over high-temperature superconductivity. Since its discovery, the ardor of many investigators has notably cooled down. The same happens when an Olympic record stays out of reach for years. But as soon as the record has been set it serves a benchmark. The possibility of producing materials with unique characteristics has been confirmed. Certainly economic considerations will effect realization of projects and it is unlikely that we will surpass the records tomorrow and make them a routine. But today we know for sure that the impossible has become accessible. And this has irreversibly changed the reference point in our attitude toward superconductivity.

Why don't superconducting transmission lines require expensive high-voltage equipment?

Chapter 28

What is magnetic resonance imaging?

> Come, come, and sit you down; you shall not budge;
> You go not till I set you up a glass
> Where you may see the inmost part of you.
>
> William Shakespeare, *"Hamlet."*

Today, there is nothing unusual about patients being given a prescription for Magnetic Resonance Imaging, or magnetic resonance tomography instead of an X-ray exam or ECG. To find out what stands behind these terms, we should start from the basics: understanding the magnetic nature of the nucleus. But, before this, we need to introduce some important concepts missing from the high school physics curriculum.

28.1   Magnetic moment

The magnetic properties of a small flat current loop placed into a magnetic field are determined by the magnetic moment of the current, which is equal to

$$\vec{\mu} = IS\vec{n},$$

where I is current, S is loop area, $\vec{n}$ is normal to a loop vector, chosen by the rule-of-thumb (see Figure 28.1). In particular, the energy of a loop in a magnetic field with induction $\vec{B}$ equals

$$W = -\left(\vec{\mu} \cdot \vec{B}\right) = -\mu_z B \qquad (28.1)$$

(z-axis is directed along $\vec{B}$). To rotate the loop to give a change of vector $\vec{\mu}$ projection from μ_z to $-\mu_z$, work $A = 2\mu_z B$ has be performed.

The electrons orbiting around the atomic nucleus may be considered equivalent to circular current and a magnetic moment can be assigned to it. The fact that electrons may have such an "orbital" magnetic moment is made apparent in the atomic energy change when the latter is placed into a magnetic field (see equation (28.1)).

Fig. 28.1: Magnetic moment of the current loop.

In the course of careful experimental data analysis, it appeared that the atomic properties of the external magnetic field were determined not only by the electron's circulation around nucleus, but also by the presence of a hidden "internal rotation" in the electron, which was named "spin". All elementary particles possesses spin (in some of them it is equal to 0). The intensity of such "rotation" is described by the spin number s, which can be only an integer or half-integer. For electrons, protons, and neutrons, $s = 1/2$. "Internal rotation", similar to orbital rotation, results in the possession of a spin magnetic moment. The projection of a spin magnetic moment to z-axis (magnetic field direction) gives values of

$$\mu_z = \gamma m_s \hbar,$$

where $\hbar$ is the Planck constant, quantum number m_s can be $(2s+1)$ integer values: $-s, -s+1, ..., s-1, s$, and γ is called gyromagnetic factor. The module of vector $\vec{\mu}$ is larger than its maximum projection : $\mu = \gamma\sqrt{s\,(s+1)}\hbar$, i.e., $\vec{\mu}$ in all stationary states is positioned at some angle with respect to the z-axis and it is rapidly spinning around this axis (see Figure 28.2). The projection of a spin magnetic moment for electrons, protons, and neutrons may take only two values: $m_s = \pm 1/2$. The gyromagnetic factor is for electrons $\gamma_e = -\frac{e}{m_e c}$, for protons $\gamma_p = 2.79 \frac{e}{m_p c}$, where m_e and m_p are the electron and proton masses, and c is the speed of light. Even neutrons possess a spin magnetic moment, despite the fact that they are electrically neutral in general[a]. The gyromagnetic factor for a neutron is $\gamma_n = -1.91 \frac{e}{m_n c}$ (neutron mass m_n is slightly larger ($\approx 3\%$) than for protons). Evidently, proton and neutron magnetic moments are three orders ($\sim 10^3$) smaller than the electron magnetic moment (their masses are approximately $2,000$ times bigger). By order of magnitude, almost all other atomic nuclei consisting of protons and neutrons should have approximately the same magnetic moment.

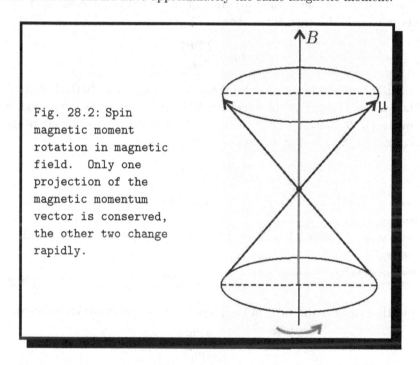

Fig. 28.2: Spin magnetic moment rotation in magnetic field. Only one projection of the magnetic momentum vector is conserved, the other two change rapidly.

Magnetic moments of nuclei were measured with great precision. It is the presence in nuclei of these tiny (as compared to atomic) magnetic moments, whose values vary for different nuclei, which is fundamental for nuclear magnetic resonance (NMR) phenomenon as well as for magnetic

[a]This testifies to the fact that neutrons should have an internal structure. Just like protons, they consist of charged quarks.

resonance tomography. We will be primarily discussing hydrogen nuclei – the protons which most frequently occur in nature. Deuterium, which is a hydrogen isotope with mass number 2, also possesses a nuclear magnetic moment.

28.2 What is nuclear magnetic resonance?

Let us consider a nucleus of a hydrogen atom (proton) placed in external magnetic field, $\vec{B}$. The proton can exist only in two stationary quantum states: in one of them, the projection of the magnetic moment upon the magnetic field direction is positive:

$$\mu_z = \frac{2.79e\hbar}{2m_p c}$$

and in the other it is negative, being the same magnitude. In the first state, the nucleus's energy in the magnetic field equals $-\mu_z B$, in the second, it equals $+\mu_z B$. Initially, all nuclei are in the first state[b]; to transfer to the second state requires an amount of energy

$$\Delta E = 2\mu_z B$$

per each nucleus. It is easy to see, then, that one can force a nucleus to change its magnetic moment projection just placing it in the electromagnetic field of frequency ω, corresponding to the transition between these states:

$$\hbar\omega = 2\mu_z B. \tag{28.2}$$

Substituting here the value of the proton's magnetic moment, one obtains

$$2\pi\hbar\nu = \frac{2.79e\hbar}{m_p c}B,$$

from where for $B = 1\,T$ one finds for the electromagnetic field frequency $\nu \approx 4 \cdot 10^7\,Hz$ and corresponding wavelength: $\lambda = c/\nu \approx 7m$ (these are the typical frequency and wavelength of broadcast band). The absorbtion by the nuclei of electromagnetic field quanta (photons) of this very wavelength results in magnetic moment flips towards the field direction. The result of

---

[b]At zero temperature, $T = 0$.

these flips is an increase in the energy of each nucleus by such a quantized amount (see Eq. (28.2)).

First of all it should be noted that in NMR experiments, i.e., for typical frequencies of medium range, the electromagnetic waves are not used in the form we are accustomed to when discussing light propagation or absorption and atomic light radiation. In the simplest case we are dealing with a coil, through which flows an alternating current created by the radio frequency generator. A sample containing the examined nuclei we want to expose to electromagnetic field effect is located along the coil axis. The coil axis is directed perpendicularly to static magnetic field B_0 (the latter is created by means of an electromagnet or superconductive solenoid). When alternating current is running across the coil, an alternate magnetic field B_1 is induced along its axis. The amplitude of B_1 is chosen much smaller than B_0 (usually 10,000 times less). This field is oscillating at the same frequency as the current, i.e., at the frequency of generator.

If the generator frequency is close to the one calculated above, intensive absorption of the radiation quanta by the hydrogen nuclei with a transition of the latter into the state with negative projection μ_z takes place. If the generator frequency is different from the calculated frequency, the absorption of quanta does not occur. Due to the sharp dependence of the electromagnetic field energy absorbtion by the atomic nuclei (accompanied by their magnetic moment flips) versus field frequency, this phenomenon was named Nuclear Magnetic Resonance (NMR).

How is it possible to notice these nuclear moment flips? It is easy, if you are equipped with the state-of-the-art NMR technology: when you turn off the radio frequency generator creating the field B_1, you should simultaneously turn on the receiver using the same coil as an antenna. In this case, it will be registering radio waves radiated by the nuclei as soon as they come back to the initial orientation along the field B_0. This signal is induced in the same coil which was used earlier to rotate the magnetic moments. Its time dependence is processed by computer and represented in the form of a corresponding spectral distribution.

From this description you can see that NMR-spectrometers differ essentially from regular spectrometers, which perform measurements in the range of visible light.

Heretofore, we were examining a simplified picture: the behavior of an isolated nucleus in a magnetic field at zero temperature. At the same time, it is clear that the nuclei are not isolated either in solid matter or in liquids.

They can interact with each other, as well as with the other excitations whose energy distribution is determined by the temperature and statistical properties of the system. Interactions between different types of excitations, their origin, and their dynamics are a subject of study of modern condensed matter physics.

28.3 How NMR was discovered

The first signals associated with nuclear magnetic resonance were obtained more than 60 years ago by the research groups of Felix Bloch[c] at Stanford and Edward Purcell[d] at Harvard University. At that time, the experimental problems were overwhelming. All the equipment was made by researchers in the laboratories on site. The exterior of the devices made in those times bears no resemblance to the modern MRI devices (with powerful super-conducting solenoids) you see at hospitals and clinics today. Suffice it to say, the magnet used in Purcell's experimental studies was manufactured from scrap metal found in the back yard of the Boston Streetcar Depot. In addition, it was calibrated so poorly that the magnetic field, in fact, had a greater magnitude than necessary for a nuclear moment flip under the radio frequency radiation $\nu = 30\,MHz$.

Purcell and his young colleagues sought the proof that the phenomenon of NMR had indeed taken place in their experiments in vain. After many days of fruitless efforts, feeling sad and disappointed, Purcell came to the conclusion that the predicted NMR phenomenon could not be observed and gave an order to shut down the power supply. While the magnetic field was decreasing, the disenchanted researchers continued to look at the oscilloscope screen, hoping to see the signals they were waiting for. At some point, the magnetic field achieved the magnitude necessary to create the resonance effect, and the NMR signal showed up on the screen.

From then on, NMR technology began moving at a quick pace. It has been widely used in scientific research in the fields of condensed matter physics, chemistry, biology, meteorology, and medicine. But its most no-table application is the imaging of the internal organs of the human body obtained with the help of NMR.

[c]Felix Bloch (1905–1983), American physicist; 1952 Nobel Prize in physics.

[d]Edward Purcell (1916–1997), American physicist; Nobel Prize 1952 in physics.

28.4 How the imaging of internal organs is performed by NMR

Until now, we were implying without saying that if the influence of the weak currents in the coils is neglected, the magnetic field where the nuclei are positioned will be homogenous, i.e., it will have the same value at every point. In 1973 Paul Lauterbur[e] suggested that NMR experiments be performed by placing a sample into an inhomogeneous magnetic field. It is obvious that in this case, the resonant frequency for the studied nuclei will change point-to-point, which makes it possible to estimate the spatial positioning of small parts of the sample. Considering that the intensity of a signal emitted from a certain spatial domain is directly proportional to the number of hydrogen atoms in this domain, one can obtain information about the matter density spatial distribution. Basically, this is what lies at the root of magnetic resonance imaging (MRI) methodology. As the reader may see, the underlying principle is simple, although, to obtain real images of internal organs, powerful computers controlling pulsed radio frequency radiation were necessary, as well as plenty of time to improve the methodology for the creation of magnetic field profiles and for processing the NMR signals received from the coils.

Let us imagine that small, water-filled spheres are positioned along the axis x (Figure 28.3). If the magnetic field does not depend on x, a single signal occurs (see Figure 28.3, a). Then, let us assume that with the extra coils (with respect to the coil creating the main magnetic field oriented along the z-axis) we create an additional magnetic field which changes along the axis x, with its magnitude growing from left to right. In this case, it is clear that for the spheres with different coordinates, the NMR signal now corresponds to different frequencies, and the measured spectrum contains five characteristic peaks (see Figure 28.3, b). The height of the peaks is proportional to the number of spheres (i.e., water mass), which have a corresponding coordinate, and thus, in the previous case the peak intensities relate as 3:1:3:1:1. If the magnetic field gradient is known (i.e., the rate of its change along the axis x), one can present the measurable frequency spectrum in the form of the hydrogen atomic density dependence on coordinate x. In this case, it is possible to say that wherever the peaks are higher, the density of hydrogen atoms will be greater: in our example

---

[e]Paul Lauterbur (1929–2007), American chemist; 2003 Nobel Prize in medicine.

the number of hydrogen atoms corresponding to the positions of spheres indeed relates as 3:1:3:1:1.

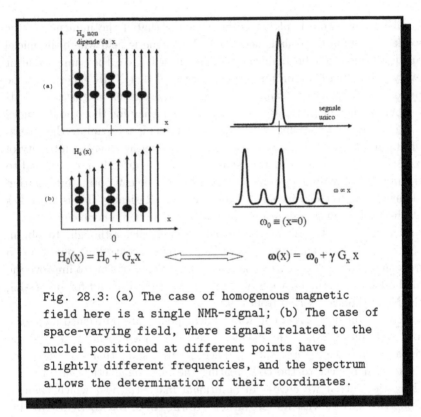

Fig. 28.3: (a) The case of homogenous magnetic field here is a single NMR-signal; (b) The case of space-varying field, where signals related to the nuclei positioned at different points have slightly different frequencies, and the spectrum allows the determination of their coordinates.

In a constant magnetic field, let us position a somewhat more complex configuration of small water-filled spheres and apply an additional magnetic field changing along all three axes. Measuring NMR radio frequency spectra and keeping in mind the values of the magnetic field gradients along the coordinates, we can create a three-dimensional map of sphere distribution (and hence, hydrogen density) in the studied configuration. This will be much more difficult to do than in the previous one-dimensional case, but the nature of the process can be understood intuitively.

An image-recovery procedure similar to the one we have described is performed during MRI tomography. When the data collection is finished, the computer begins signal "processing" with very fast algorithms and establishes a connection between the intensity of measured signals at certain frequencies and the resonating atom density at given points on the body. When this procedure is completed, a two- (or even three-dimensional) "image" of a certain organ or a body part appears on the computer screen.

28.5 Striking "images"

In order to fully appreciate the results of an MRI exam of human internal organs (e.g. different sections of brain, which today a medical physicist can obtain without even touching a cranium!), it should be understood that we are talking here about a computer recreation of "images", as opposed to real "shadows" appearing on photosensitive films, as that due to X-ray absorption during an X-ray exam.

The human eye is a sensitive meter of electromagnetic radiation in the visible range. Fortunately or not, we cannot see radiation emitted by internal organs – we can only see the human exterior. At the same time, as we have just discussed, under certain conditions the nuclei of atoms in human internal organs can emit electromagnetic waves in the radio frequency range (i.e. much smaller frequencies than necessary for visible light), and this frequency can be slightly variable depending on the radiation point. Such radiation is not visible to a human eye, therefore it is registered by sophisticated devices and then special computer processing turns it into a consolidated image. Nevertheless, we are talking about seeing the internal part of an object or of a human body in real life.

This extraordinary progress made by humankind became possible due to a number of breakthrough achievements in fundamental science: among them are quantum mechanics and its angular momentum theory, a theory of interaction between radiation and matter, digital electronics, mathematical algorithms of signal transformation, and computer technology.

The benefits offered by MRI tomography as compared to other methods of diagnostics are numerous and significant. A technician can easily choose which sectors of the patient's body to scan; he can also examine several views of a selected organ simultaneously. Among other things, proper selection of magnetic field gradients may help obtain images of vertical views of our skulls' interior (see Figure 28.4). It can obtain a central view or the view can be shifted to the right or to the left. Such examination is practically impossible to perform with X-ray tomography. Technicians can also "narrow" the observation area, visualizing the NMR signal emitted by one selected organ only, or by a certain part of it only, thus increasing image resolution. The possibility of taking direct measurements of local viscosity as well as flow direction of blood, lymph, and other body fluids is another important advantage of this exam. Selecting a required correlation between parameters, e.g. impulse time and frequency, for every condition, a technician can achieve optimum image quality, e.g., to improve the sharpness of image (see Figure 28.4).

Fig. 28.4: The image of a skull and vertebral column which, depending on the contrast dye, shows white or grey brain tissue, vertebrae, and cerebrospinal fluid with anatomic precision.

Summing up the above, one can say that for every imaging point (pixel) corresponding to a tiny volume of the object under study, it is possible to extract a variety of useful information, in some cases even the distribution of certain chemical element concentrations in the human body. To improve measurement sensitivity, i.e., to increase the ratio between signal intensity and noise, a huge number of signals should be accumulated and integrated. In this case we will obtain a quality image which adequately shows a real object. This is the reason why a MRI session takes a long time – the patient has to stay in the solenoid producing the field, motionless, for several scores of minutes.

In 1977, Peter Mansfield[f] invented the combination of magnetic field

---

[f]Peter Mansfield (1933), British physicist; 2003 Nobel Prize in medicine.

gradients which, although not providing high-imaging quality, nevertheless allowed one to receive the image very quickly: one signal is enough for appropriate plotting (in real time it takes 50 milliseconds). This technique – today it is called eco-planar imaging – allows us to trace heartbeats in real time: in this "movie", heart contractions and expansions are alternating on the screen.

Could anybody imagine in the early days of quantum mechanics that 100 years later a scientific breakthrough would make all these miracles possible?

Chapter 29

Towards quantum computers

Small is Beautiful.

E. F. Schumacher.

In Figure 29.1 you can see the famous Rosetta Stone, which was found in 1799 in Egypt, close to the town Rosetta. "In the reign of the young one – who has received the royalty from his father – lord of crowns, glorious, who has established Egypt... " was cut in a dark granite slab with dimensions of 114×72 cm^2 in honor of King Ptolemy V on the occasion of the anniversary of his coronation. Fortunately, the Rosetta Stone bears three inscriptions: the top register in Ancient Egyptian hieroglyphs; the second one, in the Egyptian demotic script; and the third one, in ancient Greek. Since ancient Greek is well known, the Rosetta Stone gave Jean-François Champollion the key to explaining the secret of Egyptian hieroglyphs in 1822.

In 1970, engineers at the Nippon Electric Company (NEC) created a dynamic memory element, which had 1,024 memory cells (small rectangles) arrayed in four grids of thirty two columns and equal number of rows (see Figure 29.2). Its size was 0.28×0.35 cm^2 and there was enough room for all the information inscribed on the Rosetta Stone. It took about 300 microseconds to upload the Rosetta Stone text on this chip.

29.1 Milestones of the computer era

The story of a radically new approach to storing and managing information began in 1936, after publication of the article "On Computable Numbers

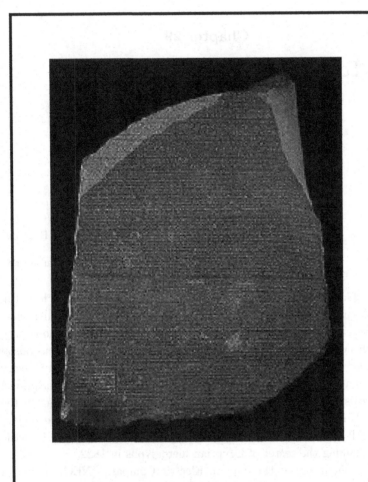

Fig. 29.1: Rosetta stone.

with an Application to the Entscheidungsproblem" by Alan Turing. The paper contained the main concepts of the logical design of a universal computer, which is now called the Turing machine.

The first generation of practically realized computers based on vacuum tubes (such as had been used in old TV sets) appeared ten years later. These computers formed the prehistory of the computer era. They were

Fig. 29.2: The first dynamic memory element.

mostly used to verify this or that theoretical concept. In essence, the first computers served as experimental facilities. The weight and the size of these "dinosaurs" were enormous; often special buildings were built for them to be placed in.

At the same time, considerable progress was achieved in electronics. In 1947, Walter Brattain[a], William Shockley[b], and John Bardeen[c] invented the first semiconductor transistor. It was realized by means of a point contact at the semiconductor substrate and had dimensions of the order of centimeters. Later, transistors were widely used in electronics as amplifiers of electromagnetic signals or controllable switches operated by electric signals. They replaced vacuum tubes worldwide.

In 1958, Jack Kilby created the first semiconductor integrated circuit (chip), which contained only two transistors. Soon after that the microcircuits integrating tens and hundreds of transistors on the same semiconductor substrate appeared.

These discoveries started up the second period of the computer era

[a]Walter Brattain (1902–1987), American physicist; 1956 Nobel Prize in physics.

[b]William Shockley (1910–1989), American physicist; 1956 Nobel Prize in physics.

[c]John Bardeen (1908–1991) American physicist; he was awarded two Nobel Prizes in physics: for the invention of the transistor in 1956 and for the creation of the theory of superconductivity in 1972 (see Chapter 25).

(1955–1964). At the same time, magnetic cores and magnetic drums (the ancestors of modern hard discs) came into use as computer data storage (often called the *storage* or *memory*).

Side by side with the rapid development of computer hardware, the ideology and architecture of logical facilities leaped forward. In June 1954, John von Neumann[d] presented the work entitled, "The first draft of a report on the EDVAC", which fully described the operation of the digital electronic computer. Moreover, in this paper von Neumann discussed in detail the fundamentals of the logic used in computer operation, and justified the use of the binary numeral system[e]. Since then computers have been recognized as a subject of scientific interest. This is why even today some scientists call computers "von Neumann machines".

All these fundamental discoveries and inventions outlined the beginning of the second period of the computer era. Computers of the third generation (1965–1974) were created on the basis of integrated circuits. More miniature and capacious semiconductor memory devices replaced bulky magnetic ones. They are still used in personal computers as the random access memory (RAM). A breakthrough in theoretical cybernetics, physics, and technology achieved in the last third of the twentieth century facilitated rapid development of computer technology allowing a tremendous decrease in computer sizes and prices making them affordable for the public. That is why since the mid-1970s computers have become an important element of everyday life. Incredible progress in technology allowed tens of millions of transistors to be placed inside an integrated device within an area of 3 cm^2 (the same area was used by J. Kilby). Modern memory devices allow a single laptop computer to store the contents of the entire US Library of Congress (150 million books and documents). The well-known Moore's law (an empirical observation made by Gordon E. Moore in 1965) in one of its forms has established exponential growth of the memory capacity and the number of transistors in a processor. One can see in Figure 29.3 how it

---

[d] John von Neumann (1903–1957) Hungarian-born American mathematician and physicist who made major contributions to a vast range of fields, including set theory, functional analysis, quantum mechanics, ergodic theory, continuous geometry, economics and game theory, computer science, numerical analysis, hydrodynamics (of explosions), and statistics, as well as many other mathematical fields. He is generally regarded as one of the greatest mathematicians in modern history.

[e] The binary numeral system, or base-2 number system, represents numeric values using two symbols: 0 and 1.

behaved over a 40-year period.

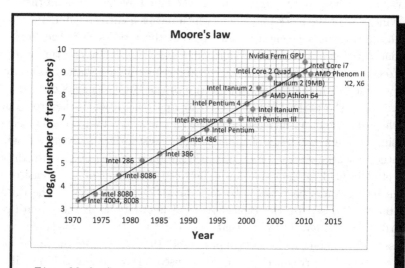

Fig. 29.3: Growth of the number of transistors in processor as a function of time (doubles every two years).

Fitting more transistors inside the same area and higher clock frequency (characterizing computer speed) became an obsession of computer manufacturers. In 2004, the Intel's Prescott processor designed with 90 nm technology (this means that several million transistors were placed on $1\,cm^2$ of its area), contained 150 million transistors and operated with the clock frequency at 3.4 GHz. In 2007, computer manufacturers switched to 65 nm technology, and soon they are going to reduce its scale down to 45 nm. Nevertheless, this straightforward method of computation power increase is almost exhausted. For example, one of the obvious restrictions is caused by the finiteness of the electromagnetic signal propagation velocity in circuits: its value depends on the technical realization of the processor, but certainly cannot exceed the speed of light $c = 300.000\ km/s$. This value seems to be enormous, but let us make a simple estimate. The clock frequency of $\nu = 3.4\ GHz$ means that the interval between two operations is of the order of $\delta t = 1/\nu \approx 3 \cdot 10^{-10}\ s$, i.e., the space separation between the units involved in two sequential operations cannot exceed $L = c\delta t \approx 10\ cm$.

Note that we obtained this result for the ideal case of signal propagation in vacuum at the speed of light! Hence, the inevitable condition of the further increase of the clock frequency (which means a corresponding increase in the number of elementary transistors in the processor) is the further miniaturization of the device. Engineers chose this path a long time ago.

Yet, we are approaching the day when further miniaturization will come into conflict with the physical properties of the basic elements of modern processors. The major scourge caused by miniaturization is the processor overheating due to an increase of released Joule heat – even when currents passing through hundreds of millions of transistors are extremely small. Another essential restriction on miniaturization is caused by the growth of the electric field in the oxide layer[f] between electrodes of the transistor: when its thickness decreases, dielectric breakdown may occur. The third limitation is posed by the typical scale of fluctuations in the impurities distribution density. These fluctuations should be able to self-average over the size of the smallest elements of processor. Hence the tinier the processor the weaker the currents that should be used. However, it is impossible to reduce the voltage to zero: beyond a certain limit the useful currents (signals) become so small that they simply get lost against the background of various noises always present in electronic circuits. This is the reason why researchers are not interested in a purposeful voltage decrease; however, they have to do it for the sake of further miniaturization. Therefore, the concept of a direct clock frequency growth has been diminished.

29.2 Twenty-first century: looking for a new paradigm

Due to the problems outlined above facing the direct clock frequency growth in traditional processors, two major manufacturers (Intel and AMD) decided in 2005 to switch to a radically new architecture of the "computer brain" – to a multicore processor. In the beginning there were two-core processors, and now we are witnessing the advent of processors with four or more cores. Such devices unify two or more independent processors while the scratch-pad memory and its controller remain common. This architecture allows a substantial increase in processing speed without an increase in the clock frequency, and thus helps to avoid the problems above. Nev-

---

[f]The insulation barrier separates the electron systems of two electrodes.

ertheless, one should not think that the two-core processor is twice as fast than the one-core one. The situation is more complex: the productivity of the multicore processor essentially depends on the type of the problem under consideration, and on the use of a brand new software, based on the advantages of parallelism. In some cases, the computation speed increases almost twice; in other cases, this is not so.

Although the development of the multicore technology is an effective engineering solution, it still remains a palliative to a clear new mainstream concept for computer technologies. It should be recognized that we have practically reached the limits of the macroworld, where the laws of classical physics still govern the behavior of electrons in the processor circuits.

This is why during the last decade scientists have been trying to work out the concepts of fundamentally new computing devices, based on new logics (different from those by Turing and von Neumann) and new basic components. Typical examples of these approaches are the development of new, *quantum*, computing algorithms (quantum computing) and of nanometer-sized devices (nanoscience, nanotechnology). To date, the idea of the realization of a "quantum computer" has resulted in the development a new interesting field of math, which, however, has still failed to provide significant practical implementations. This is why we will now review several relatively novel devices that follow the "classical" computer logic, but exploit essentially quantum phenomena.

29.3 Where does the border between the macro- and microworld lie?

As was just mentioned above, the transfer to 45 nm design rules will become the next breakthrough in computer technology. Is this scale small or large with respect to the quantum world? The characteristic scales relevant to nanoscience are presented in Figure 29.4.

The first microscopic (but still classical) length scale which we have to bear in mind while decreasing sizes of the processor's elementary blocks is the *electron mean free path*, l_e, in metal layers. Let us recall that Ohm's law is based on the *diffusive* character of electron motion when the charge transport is dominated by multiple electron scattering at impurities or other defects. Since in the process of deriving Ohm's law, the electron motion is averaged over positions of defects, the specifics of their locations is not

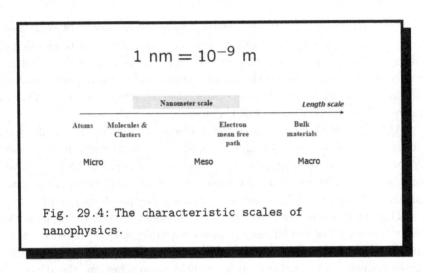

Fig. 29.4: The characteristic scales of
nanophysics.

felt in the final result. This approximation remains valid until the number
of defects in a sample becomes large. Hence, the connecting wires between
the circuit transistors can be considered as classical Ohmic resistors, when
their size L is much larger than the electron mean free path l_e: $L \gg l_e$
(in sputtered metal films l_e is of the order of several nanometers). At
the opposite limit, $L \ll l_e$, the electron motion is *ballistic* – electrons are
accelerated by the electric field over the whole device rather than diffuse
across it. One can expect that the properties of such wires would deviate
considerably from the classical Ohmic behavior.

The next important scale characterizing the limits of the quantum world
is the electron's *de Broglie wavelength*, $\lambda_F = h/p_e$. We already know that
quantization and interference are characteristic features of the quantum
world. When even the smallest dimension of a resistor becomes comparable
with λ_F, electrons obey quantum mechanics. In particular, their motion
in the corresponding direction is quantized, and the whole resistor can be
described as a quantum well. The value of λ_F depends strongly on the
electron concentration and, for a normal metal, turns out to be of the
atomic scale. However, in semiconductors λ_F can be much larger, and
the quantum confinement in corresponding nano-objects becomes crucially
important. In particular, quantum confinement along one direction allows
fabrication of a new object – two-dimensional electron gas, which serves as
a building block for many modern electronic devices.

Another purely quantum scale absent in classical physics is the so-called *phase coherence length, l_ϕ.* Let us note that an important characteristic of a quantum particle is the phase of its wave function, which is closely related to its energy. When an electron is scattered by static impurities, only the *direction* of its momentum, $\mathbf{p}_e$, changes but it does not change its absolute value, $|\mathbf{p}_e|$ or its energy. Therefore, such scattering is called *elastic*, the corresponding length scale being l_e. Even after many elastic scattering events, the electron phase remains fully determined. Hence, it can take part in quantum interference processes. However, sometimes due to a scattering event the electron spin flips, or the electron scatters by lattice vibrations with a finite energy transfer. In these cases, the coherence of the final state with the initial one is broken. The characteristic length at which the electron phase changes by 2π due to random scattering events (that means complete loss of the phase memory) is called the phase coherence length, l_ϕ. It determines a new quantum scale: when the geometrical dimensions of the device become shorter than the phase coherence length ($L < l_\phi$), quantum interference may occur in the device.

Finally, recall that the value of the electron charge in classical physics is considered infinitesimally small, so the charge flow through the capacitor plates in the process of charging or recharging is assumed to be continuous. When the device capacitance, C, is so small that the one-electron electrostatic energy, $e^2/2C$, becomes comparable to other energy scales of the system (the thermal energy $k_B T$ or the electron energy in the gate potential, eV_g) new phenomena relevant to single-electron tunneling can take place.

Table 29.1: RELEVANT LENGTH SCALES IN MESOSCOPIC DEVICES.

Conventional device	Mesoscopic device
diffusive, $L \gg l_e$	ballistic, $L \ll l_e$
incoherent, $L \gg l_\phi$	phase coherent, $L \ll l_\phi$
no size quantization, $L \gg \lambda_F$	size quantization, $L \ll \lambda_F$
no single electron charging, $\frac{e^2}{2C} \ll k_B T$	single electron charging, $\frac{e^2}{2C} \gg k_B$

Interplay between the different scales discussed above leads to a variety of regimes of quantum transport through nanosystems. The specific properties of different regimes can be used for various applications. Obviously, there is not enough room here for a detailed discussion of the physical properties of the new contenders for the roles of the basic elements for new computers, and we will briefly discuss only some of them.

29.4   Quantum wires and quantum point contacts

Let us consider how electric charge flows through the so-called quantum wire, i.e., a very thin channel which contains charge carriers but is free of impurities or other defects of the crystal lattice (see Figure 29.5).

As has already been mentioned, under the conditions when the transverse size w of a conductor becomes comparable to the electron wavelength λ_F, the electron is confined in the quantum well and its motion in the appropriate direction is quantized. This means that the part of the energy corresponding to transverse motion can take only certain discrete values, E_n, while the electron motion along the wire remains free. Since the full energy E of the electron is conserved, the larger the value of E_n, the smaller the amount of available energy left for longitudinal motion. The smaller the energy, the longer the corresponding wavelength. Hence, each energy level E_n corresponds to the specific plane wave with its characteristic wavelength λ_n (wave mode). Thus, the electron propagating in a quantum wire resembles an aggregate of wave modes in a waveguide (see Figure 29.6,a) rather than a particle diffusing in a scattering media[g] (see Figure 29.6,b).

Each mode contributes to the process of charge transfer, with the total conductivity being determined by the sum of all their contributions. The contribution of a single quantum channel to the charge transfer can be evaluated, as always, based on dimensional analysis. In an ideal case, it does not depend on the wire properties and is purely a combination of the

---

[g]Let us recall that in Chapter 4, we discussed sound wave propagation in the underwater waveguide without taking quantization into account. The fact of the matter is that we were not even considering such a possibility, while it exists. Indeed, when the sound wave wavelength is comparable to the sea depth, the underwater waveguide becomes too narrow for such waves, and only certain definite sound modes can propagate through. So the waveguide will not work well, like the Bell's water mike (Chapter 12). The characteristic wavelength must be of the order of 1 km, i.e., sound frequency $f = c/\lambda \approx 1\ Hz$. As frequency grows, the number of nodes in the standing transverse wave increases and we return to the continuous picture of Chapter 4.

Fig. 29.5: (a) Normal resistor with impurities;
(b) ballistic motion of an electron in the quantum
wire; (c) quantum point contact.

universal constants. The only available combination with the dimensionality of resistance is h/e^2, which could not appear in the classical theory. It turns out that e^2/h is the maximum conductance value, which can be realized by a single mode propagating along a quantum wire.

In Figure 29.7, one can see a real experimental realization of the quantum point contact. Its conductance at temperature $T = 1.7\,K$ changes discretely versus gate voltage V_g (which controls the effective width of the conducting channel) instead of being constant (in accordance with classical behavior). The height of each step corresponds to the value $2e^2/h$, in accordance with our speculation based on dimensional analysis.

Fig. 29.6: (a) Wave modes in quantum wire, (b) a
common resistor with impurities.

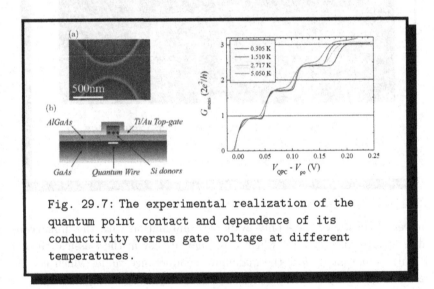

Fig. 29.7: The experimental realization of the
quantum point contact and dependence of its
conductivity versus gate voltage at different
temperatures.

The conductance quantization in point contacts usually disappears at
$L > 2 \ \mu m$; it can also be smeared out by an increase of the temperature
(see the curve corresponding to $T = 20 \ K$).

29.5 "Coulomb blockade" and single electron transistors

Let us consider properties of the so-called *"quantum dot"*: a small metallic island with size $\ll 1\ \mu m$ placed on an insulating substrate. Metal electrodes perform different functions. The first one, the "gate", changes the electrostatic potential of the dot. The second pair, the *"source"* and the *"drain"*, supply and remove electrons to and from the dot. They are arranged as shown in Figure 29.8.

Fig. 29.8: Quantum dot - small metallic drop placed on an insulating substrate with metal electrodes as shown.

Let us assume that the dot contains N excess electrons, hence its total charge is $Q = Ne$. The corresponding electrostatic energy of the dot in the absence of the external potential is $Q^2/2C = N^2e^2/2C$, where C is capacitance of the dot (usually very small). This energy is due to repulsion of the excess charges distributed over the dot surface.

When the potential V_g is applied to the gate, the total electrostatic energy of the quantum dot consists of two terms: the energy of the electrostatic repulsion of the excess electrons found above and the work of the external field on the charge transferred to the dot from infinity

$$E(N, V) = -V_g Ne + \frac{N^2 e^2}{2C}.$$

This is a quadratic function of the number of electrons, N. Its minimum is formally reached at $N = CV_g/e$, where the derivative $dE/dN = 0$.

Now let us recall that N must be an integer number. Consequently, depending on the gate voltage V_g, different situations are possible (see Figure 29.8). In the first case, when $V_g = Ne/C$, the minimum of the parabola indeed corresponds to a real state with the integer number of electrons N. However, when $V_g = (N + 1/2)e/C$, the minimum is formally realized at a semi-integer number of electrons, which is impossible. The nearest possible state to this formal energy minimum are those with the integer numbers N and $N + 1$. It is important that both states have exactly the same energy:

$$E(N = \frac{CV_g}{e} - \frac{1}{2}, V_g) = E(N = \frac{CV_g}{e} + \frac{1}{2}, V_g) = \frac{CV_g^2}{2} + \frac{e^2}{8C}.$$

We conclude that when the gate voltage is $V_g = eN/C$, the states with N and $N + 1$ electron at the dot are separated by the energy $e^2/2C$, and the energy conservation law "blocks" the charge transport through the dot (between the source and the drain). On the contrary, at $V_g = e(N + 1/2)/C$ electron transfer without any energy cost is possible, and the quantum dot is "open".

We see here that a small dot can operate as an efficient transistor. At low temperatures, $k_B T << e^2/2C$, the current flowing through such a "single electron transistor" almost vanishes at any gate voltage except specific values, $V_g(N) = e(N + 1/2)/C$. In the vicinities of these points the conductance shows sharp peaks. The experimental dependence of the quantum dot conductance (dI/dV_g) as the function of the gate voltage is presented in Figure 29.9.

One can hope that in the future, single electron transistors will allow development of logical circuits operating with the smallest possible currents and having extremely low dissipation.

29.6 Concluding remarks

We presented here a very brief, superficial, and far from complete review of the fascinating new research area of nanoscience, which has embraced a

Fig. 29.9: Experimentally measured dependence of
the differential conductance of the quantum dot
versus gate voltage.

lot of findings from physics and chemistry. Let us list other possible candidates for the building blocks of nanodevices, which belong to the front end
of the present research. These are: graphene[h] – a single layer of graphite,
where the electrons form two-dimensional electron gas with the relativistic
spectrum; molecular devices such as carbon and non-carbon nanotubes and
nanoelectromechanical systems; devices for quantum computation based on
various principles; spintronic devices based on manipulation of individual

---

[h]Graphene is a thin flake of ordinary carbon, just one atom thick. A. Geim and
K. Novoselov have shown that carbon in such a flat form has exceptional properties
that originate from the remarkable world of quantum physics. As a material, graphene
is completely new – not only the thinnest ever but also the strongest. As a conductor
of electricity, it performs as well as copper. As a conductor of heat, it outperforms all
other known materials. It is almost completely transparent, yet so dense that not even
helium, the smallest gas atom, can pass through it.

Andre Geim (1958), Dutch physicist; director of the Manchester Centre for Meso-
science and Nanotechnology at the University of Manchester.

Konstantin Novoselov (1974), British–Russian physicist; researcher at the same centre.

Both were awarded the 2010 Nobel Prize in physics "for groundbreaking experiments
regarding the two-dimensional material graphene".

spins; superconductivity and magnetism at nanoscale, etc. Nanoscale systems require an understanding of the peculiarities of quantum transport and the interplay between electron–electron interaction and disorder, as well as the role of contacts and the electromagnetic environment of nanodevices. All these issues are far from being fully understood yet.

Afterword

Little by little our tale about physics comes to the end. We told you how physics helps to explain so many things all around us. Remember meandering rivers and the blue sky, think of coalescing droplets and hissing tea-kettles, don't forget the singing violin and the chime of goblets. Still the magic of physics is not solely the power to explain what happens, but the ability to foresee what will happen even if it never has before. This has put physics in the forefront in scientific and technical progress.

Modern physics has opened the way to the amazing quantum world. There, prisoners of potential wells flee away from their dungeons like the Count of Monte Cristo; magnetic fields make vortices to pierce superconductors; and a volatile amalgam of wave and particle entities of light quanta bring to mind mythical centaurs. The wonders of the quantum world are beyond imagination. But using its mathematical arsenal, theoretical physics succeeds in describing behavior of quanta so accurately that the results of experiments exactly coincide with theoretical predictions. This capability to correctly represent phenomena which escape even mental visualization is, in the opinion of the world-renowned physicist L. D. Landau, the greatest triumph of theoretical physics of the twentieth century.

Index

About the Authors

Lev Aslamazov (1944–1986)

 Lev Aslamazov was born January, 6, 1944 in Batumi (Georgia). In 1967 he achieved his Master's degree *cum laude* at the Moscow Institute for Physics and Technology. In 1970, under supervision of Anatoly Larkin, he got his Ph.D. in Condensed Matter Physics at the Landau Institute for Theoretical Physics. From 1970 he worked in the same institute as a researcher, and then senior researcher, before becoming an associate professor (1976–1980) of the Moscow Institute for Steel and Alloys (Technological University) in 1976. In 1980 he was awarded the degree of Doctor of Sciences (Habilitatus) in Condensed Matter Physics and became a full professor of the Department of Theoretical Physics of the same university. His main fields of scientific interests were: superconductivity, theory of metals, and theory of phase transitions. He authored classical works in superconductivity and three books. For many years he executed the duties of vice-Editor in Chief of the popular science magazine of the Russian Academy of Sciences "Kvant".

Andrey Varlamov

Andrey Varlamov was born April, 25, 1954 in Kiev (Ukraine). In 1977 he achieved his Master's degree *cum laude* at the Moscow Institute for Physics and Technology. In 1980, under supervision of Alex Abrikosov, he was awarded his Ph.D. in Condensed Matter Physics. He worked as a researcher (1981–1985), associate professor (1986–1989), and full professor (1990–1999) of the Department of Theoretical Physics of the Moscow Institute for Steel and Alloys (Technological University). He worked as an invited fellow at the Condensed Matter Theory Group of Argonne National Laboratory, USA (1993). Since 1999 he has been the principal investigator of the Institute of Superconductivity, Innovative Materials, and Devices of the Italian National Research Council (CNR). Andrey Varlamov was awarded the USSR State Prize in Physics for young scientists (1986), the degree of Doctor of Sciences (Habilitatus) in Condensed Matter Physics (1988), and the degree of Doctor Philosophiae Honoris Causa of Bogolyubov Institute for Theoretical Physics (2011); he is also a member-correspondent of Istituto Lombardo Accademia di Scienze e Lettere (2009). His main fields of scientific interests are: superconductivity, theory of metals, theory of phase transitions, and nanophysics. The author of eight books, five monographic review articles, and more than 120 scientific papers, Andrey Varlamov is also a member of the Editorial Board of the Russian Academy of Sciences' popular science magazine "Kvant". For many years he executed the duties of vice-Editor in Chief of this magazine (1986–1992).